South Wales Valleys

Railways & Industry on the BRECON & MERTHYR

MERTHYR-PONTSTICILL JUNCTION-BRECON

Front Cover Image: A classic scene at Brecon as Brecon-based 2251 Class 0-6-0 2235 has the 2.15pm service to Newport. (W.A. Camwell/SLS)

Half title page: Pontsticill on a warm and sunny day was a pleasant place to be alongside the reservoir, as here on 9 June 1962 with Ebbw's 2218 on a service from Brecon to Newport. (Colour Rail)

Back Cover Images:
The 6.15pm Brecon to Newport at Pontsticill behind Ebbw Junction's 3634 0n 22nd August 1957, conveying a milk tanker which ran daily from Builth Wells to Aberdare which will be detached for onward transit via Merthyr. (Ian L. Wright)

The very detailed station nameboard at Cefn Coed on the joint B&M/LNWR line. (Ken Mumford Presentations)

3747 gets up steam at the Pentir Rhiw stop to tackle the Severn Mile Bank with a Newport to Brecon service on 7th September 1962. (Michael Roach)

South Wales Valleys

Railways & Industry on the BRECON & MERTHYR

MERTHYR-PONTSTICILL JUNCTION-BRECON

JOHN HODGE & RAY CASTON

AN IMPRINT OF PEN & SWORD BOOKS LTD.
YORKSHIRE – PHILADELPHIA

First published in Great Britain in 2023 by
Pen and Sword Transport
An imprint of
Pen & Sword Books Ltd.
Yorkshire - Philadelphia

Copyright © John Hodge and Ray Caston, 2023

ISBN 978 1 39904 108 9

The right of John Hodge and Ray Caston to be identified as authors of this work has been asserted by them in accordance with the Copyright, Designs and Patents Act 1988.

A CIP catalogue record for this book is available from the British Library.

All rights reserved. No part of this book may be reproduced or transmitted in any form or by any means, electronic or mechanical including photocopying, recording or by any information storage and retrieval system, without permission from the Publisher in writing.

Typeset in Palatino 11/13 by SJmagic DESIGN SERVICES, India.

Printed and bound by Printworks Global Ltd, London/Hong Kong.

Pen & Sword Books Ltd incorporates the imprints of Pen & Sword Books Archaeology, Atlas, Aviation, Battleground, Discovery, Family History, History, Maritime, Military, Naval, Politics, Railways, Select, Transport, True Crime, Fiction, Frontline Books, Leo Cooper, Praetorian Press, Seaforth Publishing, Wharncliffe and White Owl.

For a complete list of Pen & Sword titles please contact

PEN & SWORD BOOKS LIMITED
47 Church Street, Barnsley, South Yorkshire, S70 2AS, England
E-mail: enquiries@pen-and-sword.co.uk
Website: www.pen-and-sword.co.uk

or

PEN AND SWORD BOOKS
1950 Lawrence Rd, Havertown, PA 19083, USA
E-mail: Uspen-and-sword@casematepublishers.com
Website: www.penandswordbooks.com

CONTENTS

Dedication 6

Acknowledgements 6

Introduction 7

Chapter 1   Services To & From Brecon 8

Chapter 2   Location Analysis 14
- Merthyr High Street 14
- Rhydycar Junction 21
- Cyfarthfa Iron Works 24
- Ynysfach Ironworks 28
- Heolgerrig Halt 28
- Llwyncelyn Junction 30
- Cefn Coed 32
- Vaynor Siding 40
- Pontsarn 40
- Morlais Junction 46
- Pontsticill Junction 48
- Dolygaer 73
- Torpantau 80
- Pentir Rhiw 105
- Talybont-On-Usk 118
- Talyllyn East Junction 132
- Talyllyn Junction 135
- Groesffordd Halt 153
- Brecon Yard & Shed 155
- Brecon Station (Free Street) 168
- Brecon Mountain Railway 183

DEDICATION

The first volume in the series was dedicated to Ray's father, who worked on the B&M all his working career. This volume is dedicated to my father, Arthur Percival Hodge, known to all as Perce and Percy, who the late 1940s and early 1950s worked for the Western Welsh Omnibus Co. as a driver and was one of a small group who were entrusted with driving parties to points of interest and beauty throughout South Wales for day trips. He often visited part of the area covered by this volume, when he drove summer parties to the reservoirs, especially Taf Fechan at Pontsticill, and Dolygaer. As a former driver for the company at their Cross Keys depots, he had built up a useful road knowledge of South and Mid-Wales from similar outings, and on his transfer to the Barry depot, this was well received. In railway parlance he 'knew the road', something well valued in the transport world. After leaving the Western Welsh, he was employed looking after Dinas Powys railway station, which involved learning all the new language and practices of railways. He finally moved into the Ambulance Service, first as a driver, rising to become an assistant Superintendent for the Barry depot. Following his retirement in 1965, he became a Town Councillor but was dogged by ill health from 1970 and died in 1974 at the age of 73.

ACKNOWLEDGEMENTS

Many railway photographers visited this very photographable area, especially in its final years before closure and we have used photographs from a wide variety of photographers, to whom our grateful thanks are extended. In the early 1920s, a local Merthyr photographer Angus Lewis recorded the local scene as the privately owned railways in the Merthyr area went into government-sponsored amalgamation into the Great Western Railway. The Brecon and Merthyr appears to have been one of Angus' favourites and he took hundreds of pictures of B&M trains and engines at all the prime locations between Merthyr and Pontsticill, affording a superb record of railway activity in the Merthyr area in the early 1920s. His collection became part of the wider Ken Nunn collection and was then taken over by the Locomotive Club of Great Britain and has recently passed to the National Railway Museum.

We are as usual very grateful to R.A. (Tony) Cooke for use of his Track Layout Diagrams in the area and Colliery Gazetteer in the area covered by this book and also to Ian Reese of Newport for his help in editing.

Every effort has been made to establish the correct ownership of photographs and other material used in this book, but if we have failed to properly identify any images etc., please contact john_hodge@tiscali.co.uk if there is a need to rectify the position.

INTRODUCTION

This is the third in our series of books on the Brecon & Merthyr and brings to an end our study of this fascinating line and its branches. Please see the first volume for a History of the Brecon & Merthyr Company and also the second volume for an analysis of the Freight Services at the Grouping.

The railway between Pontsticill and Brecon was set in some of the most beautiful and dramatic scenery in the country and gave rise to the saying that it was 'better to travel on the Brecon & Merthyr than to arrive', especially if you were travelling to one of the more industrialised towns on the southern part of their system. To the west of the railway, north of Pontsticill, large reservoirs had been constructed to provide water to the South Wales Valleys and these formed a lovely spectacle to the traveller with the mountains of the Brecon Beacons providing a contrasting view on the east side. Added to this, such lovely villages as Torpantau and Talybont-on-Usk provided facilities for days out. The area was not without its problems for operating a railway. Principal to these was the Seven Mile Bank between Torpantau and Pentir Rhiw, seven miles at 1 in 38, which demanded a full water tank and a skilful fireman on the engine, which was limited to 120tons, and required double heading of many freight trains taking coal and agricultural traffic to historic Brecon.

The section between Merthyr High Street station and Pontsticill was much photographed at the Grouping in and around 1922 by local photographer Angus Lewis. The best of Angus Lewis' photographs were used in John Hodge's book *Six Railways to Merthyr*, published in 2014 by the Welsh Railways Research Circle and now out of print. As this work is no longer available, we have used several of those photographs featuring the B&M line from Merthyr to Pontsticill in this book. This will afford a view of the Merthyr-Pontsticill section both at the Grouping and in the years between then and closure.

Chapter 1

SERVICES TO & FROM BRECON

The biggest single event which affected services on the B&M at the north end between Merthyr-Pontsticill-Brecon, was the closure of the Cyfarthfa Ironworks at Merthyr and Dowlais Ironworks. Though Cyfarthfa re-opened briefly in the First World War, this brought to an end the period of Merthyr's greatness in that field. The aftermath began the end of its period of dominance and the start of its recession which was to last throughout the period until the Second World War and beyond, as all services have now disappeared except for that to Cardiff and Barry Island, handled over a single line in the Merthyr area. The only positive thing that can be said of the last sixty or so years is that it survived the Beeching cuts of 1964, which its sister town Aberdare initially did not, though its services were later restored. By far the biggest change in services occurred over the Merthyr-Brecon section, as the through services were progressively eliminated, cut short at Pontsticill where connections were made into the through services to and from Newport and Brecon. The latter were never many in number and remained constant from the early days of the service right through until the end with three or four services each day, the times bearing a remarkable similarity. A huge loss was also the through Summer services to Aberystwyth, the Taff Vale/Rhymney/B&M/Cambrian sponsored trains which originated at Cardiff Parade, Treherbert, and in earlier days at Newport and also the GWR-introduced Barry to Llandrindod Wells, all of which ran via Merthyr. The Aberystwyth services changed engines in pre-Grouping days from Taff Vale to B&M double-headed tank engines or Cambrian 0-6-0s and post-Grouping to GWR 2301 Class Dean Goods. These services were discontinued at the onset of the 1939 war, never to re-appear.

In 1910, the Midland Railway provided an early morning train from Brecon to Swansea via the Neath & Brecon line, and two trains from Hereford to Swansea, interspersed with two workings from Hereford to Brecon. In the Up direction, there was an early train from Brecon to Hereford, followed by three trains from Swansea to Hereford.

The Cambrian provided three trains from Moat Lane to Brecon and three from Brecon to Moat Lane, with two from Builth Road to Brecon and one in return. The B&M itself provided three trains to and from Newport and one working to Dowlais and return.

In 1922, the Midland workings were as in 1910; the Cambrian provided four trains from Moat Lane to Brecon with three in return, plus one from Brecon to Builth Road. The B&M now provided four trains to and from Newport with appropriate connections.

The 1923 Service Timetable shows an 8.25am Newport to Brecon (the later 8.3am), 10.45am Newport to Pontsticill (connecting into the 12.10pm Merthyr to Brecon) (later the 11.15am Newport to Brecon), 2.45pm Newport to Brecon (later 3pm), and 6.40pm Newport to Brecon (later 6.55pm, finally 7.7pm). Thus, over forty years the service at the Grouping was still recognisable when the line closed at the end of 1962. A post-Grouping

development by the GWR was an express service from Merthyr to Newport via the former Pontypridd, Caerphilly and Newport line, also worked by GWR 2301 Class. This was supplemented by a 'motor-car' service, which called at the intermediate halts between Pontypridd, Caerphilly and Machen. This latter service survived until withdrawn in 1956, the express service being an early casualty of the 1939 war.

In 1930, the LMS withdrew the through passenger and freight workings to and from Swansea. They continued to provide

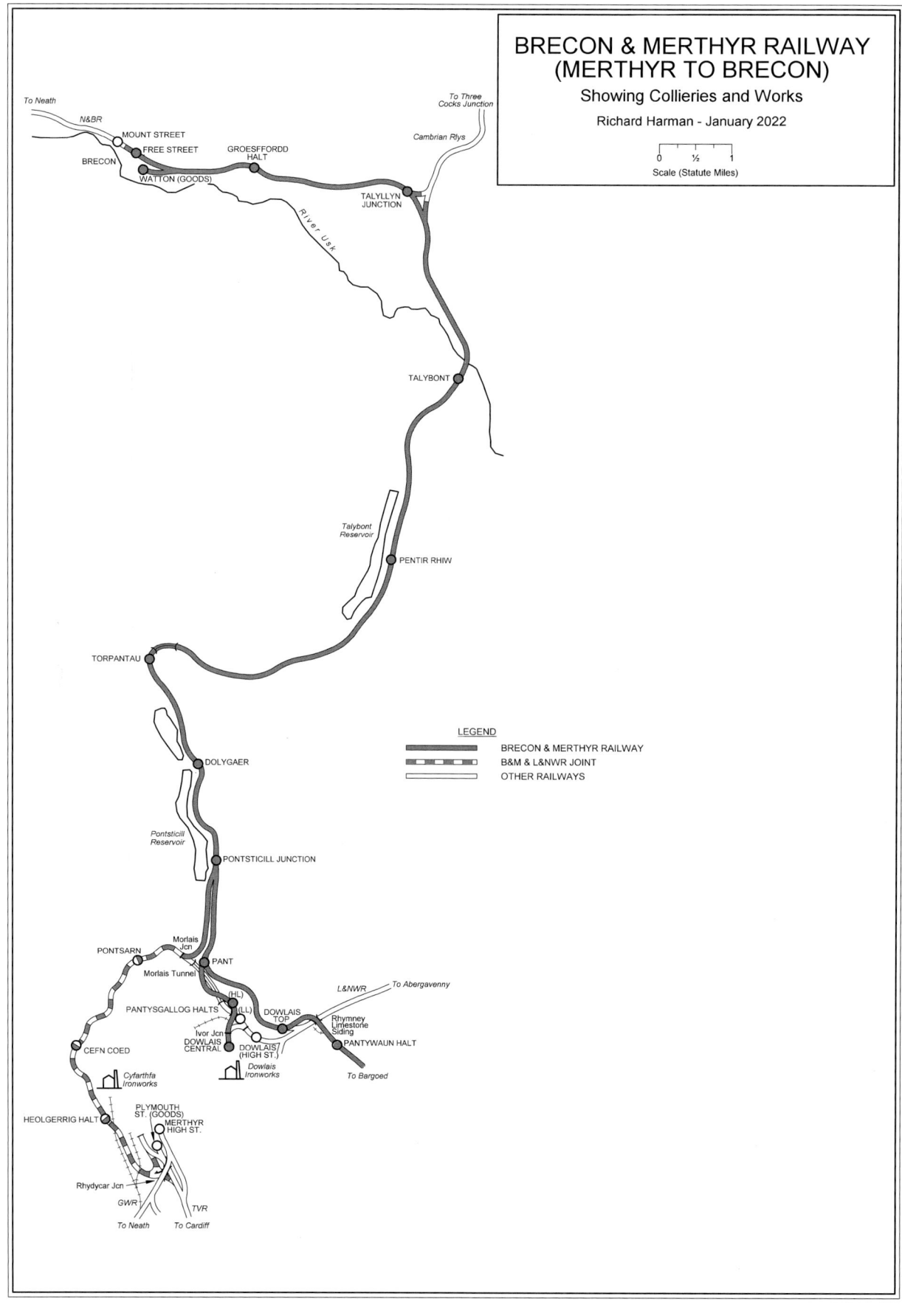

passenger and freight services to and from Hereford and Brecon whilst the GWR provided a passenger service from Neath (Riverside) to Brecon and had to institute a daily local freight service from Brecon to Colbren Jct. and return.

In 1938, the LMS provided four trains in each direction between Hereford and Brecon. The Mid-Wales line had four trains from Moat Lane to Brecon with three in the opposite direction. In each direction, there was a service with a long wait at Builth Wells. The ex-B&M service remained at four trains per day to and from Newport, whilst on the former Neath & Brecon, there were three trains in each direction, though by September 1952, the late morning services in each direction had become Saturdays Only.

The through services between Merthyr and Brecon and reverse were soon consigned to history by the GWR after the Grouping when they replaced them with connecting services between Merthyr and Pontsticill Junction. The Merthyr to Swansea services were replaced by an auto service from Merthyr to Hirwaun where they connected into Vale of Neath services between Pontypool Road and Neath and reverse, services to Swansea (East Dock) having ceased in 1935.

It was the practice for Brecon-bound trains to run into the Up back platform at Bargoed, at the north end of which they took water for the climb to Torpantau. Whilst the line and stations to Deri Junction were Rhymney Railway property, in pre-Grouping days, the ordinary passenger service was provided by the B&M's usually four trains per day in each direction. The RR did exercise its running powers over the B&M to Fochriw for coal traffic, and workmen's trains were run to suit the shifts at local collieries. Prior to the demise of iron making at Dowlais ironworks, there was a huge coalfield in the area of Pantywaun, with trains both running locally to and from Dowlais with others coming in from the Rhymney and Bargoed directions. If ever an area changed beyond all recognition, it was certainly Pantywaun, which was destroyed when it became part of Cwmbargoed Opencast. The short single platform that survived until the line closed gave no idea whatever of its earlier history.

Post-Grouping, the GWR gradually began to supplement the basic passenger service by introducing trains Bargoed to Fochriw (then a very important mining area) and Dowlais Top from where a connecting bus service to Merthyr operated. These expanded to provide a service for those working in Cardiff and there was a shopping service on Saturdays to Cardiff Queen Street, this continuing until line closure.

Perusal of the relevant working timetables of the later 1950s leads one to the conclusion that closure was already contemplated as there was little in the way of positive service development, and timings and provision were becoming more and more unattractive. Management would doubtless protest that they were being pressurised by the rise in private car ownership and bus service expansion, but the rail services provided were often not useful or conducive to potential travellers. A prime example was the service between Neath and Brecon, where on Saturdays Excepted there was a single late afternoon service from Neath to Brecon and return. The service on the Mid-Wales line was not much better, where it was possible to travel from Brecon to Builth Road in the morning, but the first train to Moat Lane did not leave until 1.20pm. The only later train left at 5.6pm but did not run beyond Builth Road on Saturdays. On the former B&M section, there was no service to Newport SX after the 2.5pm., though a 6.15pm (the return of the 3pm from Newport) was added later.

It was therefore no great surprise when proposals appeared to withdraw all the passenger services to and from Brecon in the early part of 1962, thus pre-Beeching. It took some time to arrange replacement bus services in some areas, but withdrawal of rail passenger services was completed to plan in December 1962. One wonders whether a Welsh Assembly, had it existed in those days, would have agreed with such a procedure. This left a gradually declining freight service from Merthyr to Brecon which ceased in May 1964. A commemorative last train was able to run on 2 May 1964 by the Stephenson Locomotive Society from Merthyr to Brecon and Dowlais Central on the return.

Much of the former Northern Section of the B&M then lay amongst the weeds awaiting the demolition and recovery teams. It was not until the autumn of 1965 that the track lifters came to attack the Seven Mile Bank. It was soon found that the 350HP diesel shunters being employed could not cope with the gradient and what are now Class 37 diesels were brought in to replace them. This was the only use of diesel power on the Northern section apart from visits of ex-GWR railcars on enthusiast specials.

Passenger Services

In B&M as later in GWR days, connections were normally made at Pontsticill into the Newport to Brecon and return services, though there was a through train from Merthyr to Brecon at 12.10pm (arriving at 1.40pm) and one at 2.20pm on ThSO (arriving at 4.35pm). Services to Pontsticill were at 9.35am, 2.50pm (through train to Newport) 5.35pm and 7.20pm. A notable feature pre-1914 was a 2.5pm service from Brecon double-headed with portions for Newport and Merthyr, which split at Pontsticill and attached the 2.50pm from Merthyr. This uneven pattern of passenger arrivals and departures at Merthyr led to train crews and engines working both passenger and freight trains in the course of a shift. This continued on parts of the former B&M until withdrawal of the passenger services.

The GWR continued the pattern of connecting with the four through trains between Newport and Brecon in both directions at Pontsticill but by Nationalisation, not all the Down trains from Brecon had onwards connections. Declining patronage enabled Merthyr services to be formed with just a single autocoach, often hauled by a non-auto-fitted pannier tank, utilising run round facilities at both ends of the journey. In the event, passenger services between Merthyr and Pontsticill were withdrawn earlier than on the main line, ending on 13 November 1961.

Goods Services

Loads were severely limited by the seven miles of 1 in 38 (Seven Mile Bank) up to Torpamtau Tunnel and the 1 in 45 ruling gradient from Merthyr to Pontsticill. The 1902 WTT shows a 'Talyllyn Day Goods' and a 'Talyllyn Night Goods', both double-headed to Pontsticill with a maximum of around twenty wagons per train. These would then separately work trains down to Merthyr (Taff Vale), apart from one of the engines of the Day Goods which worked a passenger train from Pontsticill to Merthyr. Loads back from Brecon to Pontsticill would again be limited to around fourteen wagons per train. There was also a working from Brecon that proceeded in much the same way, except that it worked to Pontsticill and back to Brecon, before running again to Merthyr. The upshot was that five trains worked to the Taff Vale and two to the GWR at Merthyr. Transfer trips between the Taff Vale sidings and the GWR would be worked by B&M engines as required, all part of the imbalance between passenger and goods services, reflecting on the engine and traincrew workings. The 1938 WTT shows an unbalanced Merthyr to Moat Lane goods with a Merthyr to Talyllyn and a Merthyr-Brecon, both with returns. By 1955, this had changed to two freights each from Merthyr to Talyllyn Jct. and to Pontsticill Jct. At least one of those to Talyllyn was regularly double-headed, the WTT indicating that assistance might be necessary on the return working of the afternoon train from Talyllyn if worked by one engine. After closure of the passenger service, the declining freight service continued to run to Brecon, largely composed of domestic coal and agricultural traffic with returning empties, until final closure in May 1964.

CHAPTER 2
LOCATION ANALYSIS

Merthyr High Street

Merthyr High Street station was opened on 2 November 1853 by the then broad-gauge Vale of Neath Railway, after surmounting considerable difficulties in the construction of Merthyr Tunnel on their line from Swansea. The station was typical of the Brunel design with arrival and departure platforms separated by two carriage sidings and an overall timber roof at the buffer stops end in train shed fashion.

By early 1865, a third rail had been laid down to accommodate narrow gauge working from the Pontypool direction, allowing the Vale of Neath to operate on a mixed gauge basis, the broad gauge finally being abolished throughout South Wales in April 1872. The availability of the third line facilitated the entry of the B&M into Merthyr on 1 August 1868. After the removal of the broad gauge, there followed the removal of the Taff Vale passenger trains from their original Plymouth Street station (opened in April 1841) to High Street on 1 August 1877, and the entry of the LNWR from Dowlais and Abergavenny in June 1879. The final arrival was that of trains on the Quakers Yard & Merthyr Joint Railway, opened by the GW and Rhymney railways in April 1886. This covered the period of Merthyr's

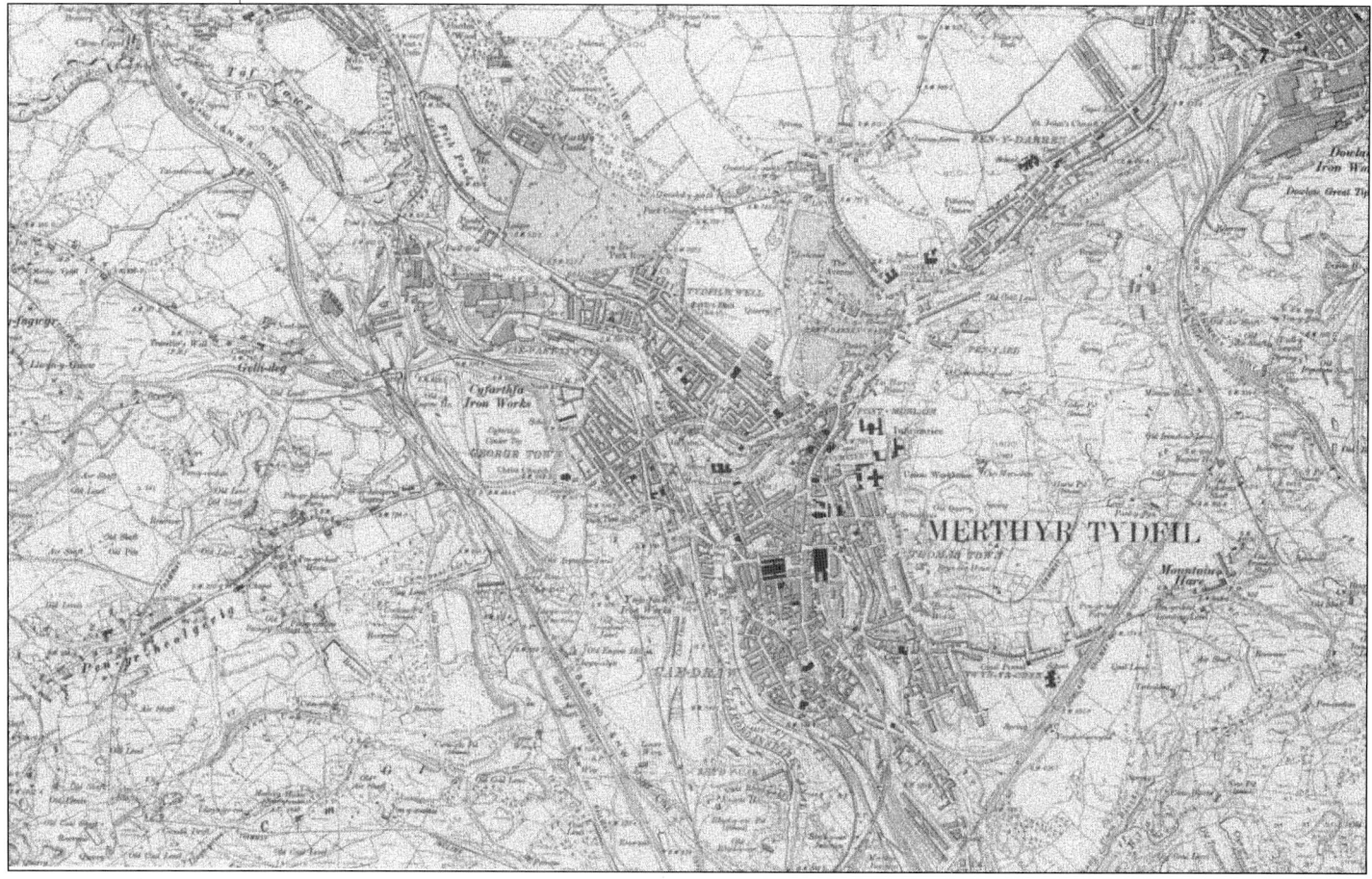

OS Map of Merthyr 1888-1913. (National Library of Scotland)

industrial greatness when it was the most important town in South Wales, due to the importance of its iron making industry alongside that of Dowlais.

Such a level of activity was beyond the capability of a two platform station and by the late nineteenth century, Merthyr had become a five platform station, by the introduction of a rather narrow wooden island platform in place of the former central carriage sidings, plus a platform and carriage siding to the outside of the overall roof on the Down side of the station. Beyond this on the Down side was a large goods yard and engine shed complex, stretching beyond the station signal box to the next box at Mardy Jct. The Taff Vale confined its goods activity to their Plymouth Street depot but stationed its locomotives at High Street joint depot.

In January 1894, there were around twenty-eight passenger services departing daily with about twenty goods workings, not including local transfers. By 1901, daily passenger train departures had risen to thirty-two and by 1910 to over fifty. Following the Grouping, the number of passenger workings had risen due to the introduction of a Merthyr to Newport and return service with around eighteen goods workings. Prior to the Second World War, there were still around fifty passenger workings each day, including through services from Cardiff/Treherbert to Aberystwyth and Barry to Llandrindod Wells, each of which reversed and changed engines at Merthyr, the former bringing Cambrian 0-6-0 engines to Merthyr in some years, and in others being worked by B&M double headed tank engines. Following the Grouping, these passed to GW working with double headed veteran panniers and 2301 Class 0-6-0s.

From September 1939, the longer distance passenger workings to Mid-Wales (by then to Moat Lane) and Newport

Merthyr showing part of the original layout in the 1920s with an unrebuilt Taff Vale Class A departing with a service to Cardiff. (LCGB/NRM)

The large B&M 0-6-2 Saddle Tanks were much used on Merthyr to Brecon services and here No. 26 on the 12.10pm service stands alongside LNW stock on an Abergavenny train at Merthyr station c.1922. Unfortunately, these engines did not last long under the GWR and all were withdrawn shortly after the grouping, the final one in 1928. (LCGB/NRM)

were withdrawn and in 1951 workings over the Quakers Yard & Merthyr Joint line were curtailed by the deterioration of Quakers Yard viaduct just outside the High Level station there. In fairly rapid succession, the services to Abergavenny ceased in January 1958, to Pontsticill Jct. in November 1961 and Hirwaun in December 1962, leaving only the service to Pontypridd, Cardiff and Barry, which fortunately survived the Beeching cuts of 1964. With the national withdrawal from sundries traffic in 1972 and wagon load in 1976, all by then handled at Plymouth Street, this left only the hourly DMU passenger service which deteriorated in use during the later 1960s and 70s until rescued by the regeneration of the Valleys services during the 1980s. The facilities required for this service were far too great as provided by the former layout and this was gradually reduced until a new single platform was created with the former train shed sold off for redevelopment as a Tesco store. The station is now managed by Transport for Wales. Just over half a million passengers a year use this service, though the level has reduced in the last five years; this however should be a firm basis for a positive development for the future.

John Hodge's book *Six Railways to Merthyr* recounts the industrial and social history of the town leading up to the Grouping and beyond and is recommended to those who wish to appreciate the railway photographs of Angus Lewis in the early 1920s.

There was a service each day at 2.50pm from Merthyr to Pontsticill in B&M days which on arrival was attached to a Brecon to Newport train. The train from Merthyr was always worked by a Dowlais engine 17 or 18 with target D2 and is seen here waiting to depart at Merthyr c.1922. (LCGB/NRM)

Summer services ran from Cardiff and Treherbert to Aberystwyth joining up at Pontypridd and worked from Merthyr by Cambrian 0-6-0s as here with 885 c.1922. Some years the B&M worked these trains with two engines on from Merthyr and after the Grouping the GWR worked them with two Panniers or a 2301. These trains took the Talyllyn East Junction curve to access the Mid Wales lines. (LCGB/NRM)

Two B&M saddle tanks are seen backing onto an Aberystwyth train at Merthyr c.1922 and will work the train to Talyllyn East Junction where a Cambrian engine will take over. (LCGB/NRM)

B&M 8 on the Ynysfach Loop at Merthyr with a B&M Brakevan with the Taff Vale line to Plymouth Street in the background and the GWR line into High Street on the Viaduct, c.1922. This engine fared quite well under the GWR and lasted until 1933 as 2184, based at Ebbw Junction. (LCGB/NRM)

LOCATION ANALYSIS • 19

9618 with the 4 coach 10.02am Pontsticill to Merthyr passenger at Merthyr on a wet morning in November 1961. (W.G. Sumner)

The interior of the Train Shed at Merthyr, Platform 1 to the right, 2/3 centre, 4 to the left.

Present day Merthyr with a Class 150 DMU with a service to Bridgend via Barry and the Vale of Glamorgan. (Michael J. Back)

Map of the Merthyr and Dowlais areas, showing the B&M lines.

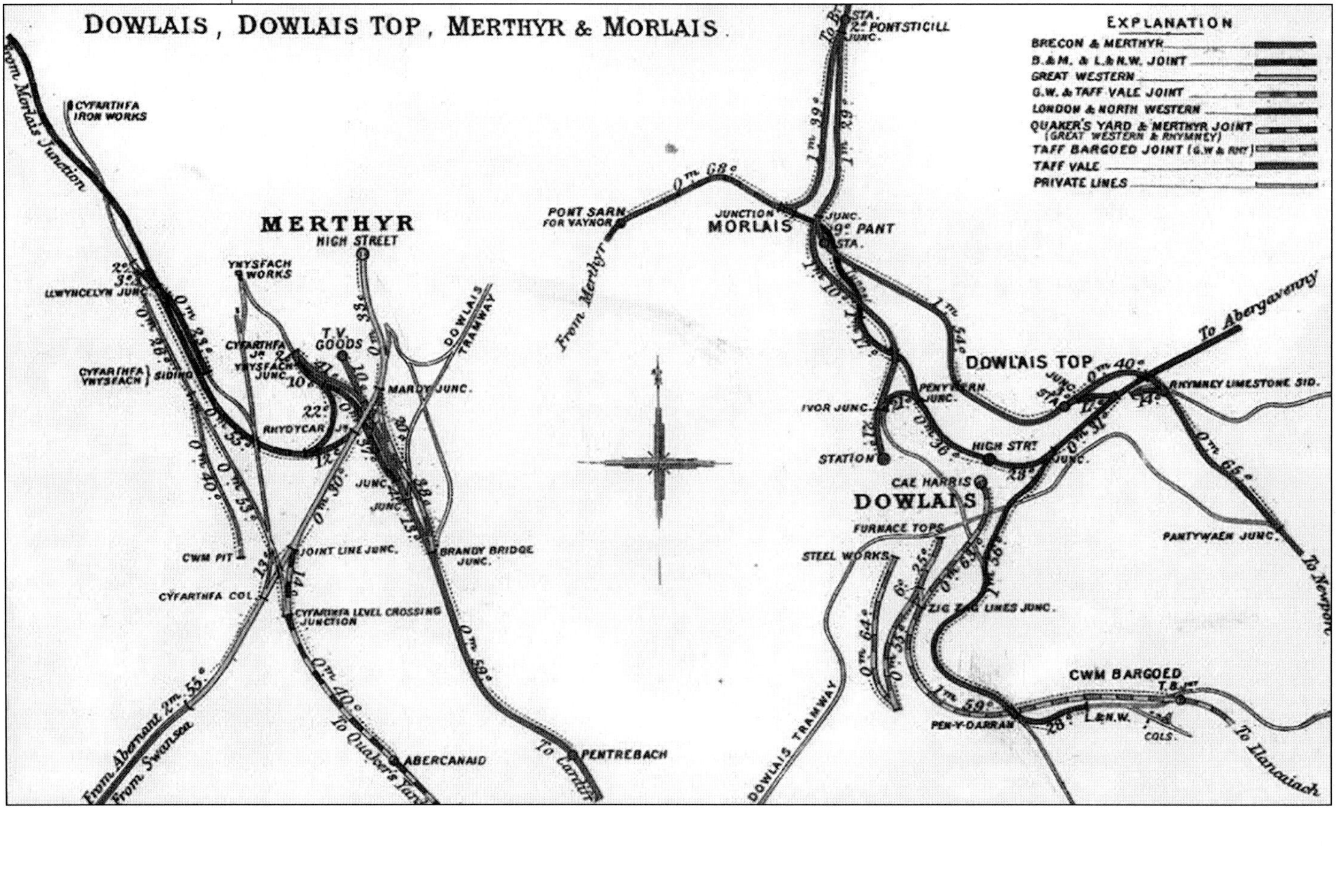

Rhydycar Junction

Rhydycar Junction at 23miles 56ch from Brecon was the junction between the Vale of Neath line to Swansea and that to Pontsticill. Approaching Rhydycar Jct. down the joint line from Cefn Coed, another very important junction was reached first, Ynysfach Jct., the junction for the Ynysfach Loop. Originally this had been built to connect the Ynysfach Ironworks to the Taff Vale Railway, but the B&M acquired the relevant take-over powers in its Act of 1868. By means of a reversal, the 'High-Level Loop' connected directly with the TV Sidings without having to access the GWR. The importance of this connection was apparent in the number of B&M freight trains that terminated here. At Rhydycar Jct. proper, the Vale of Neath line, by then part of the GWR, was reached giving access to Merthyr High Street station and yards, at 24miles 20ch from Brecon.

Rhydycar Junction, showing the former B&M/LNWR line curving right towards Cefn Coed, the former Vale of Neath line runs straight on towards Merthyr Tunnel. (Ken Mumford Presentations)

A Cardiff/Treherbert to Aberystwyth service via Merthyr rounds the curve at Rhydycar Junction onto the B&M/LNW Joint line to Morlais Jct. c.1922 with B&M 2 piloting an 0-6-2 Saddle tank for the run to Talyllyn. (LCGB/NRM)

The complications of Rhydycar Jct. are well shown on this extract from Tony Cooke's 1947 GWR Atlas. This is the situation after the remodelling of 1936 after Cyfarthfa Crossing SB had been taken out and the lines to Merthyr Tunnel and Abercanaid singled.

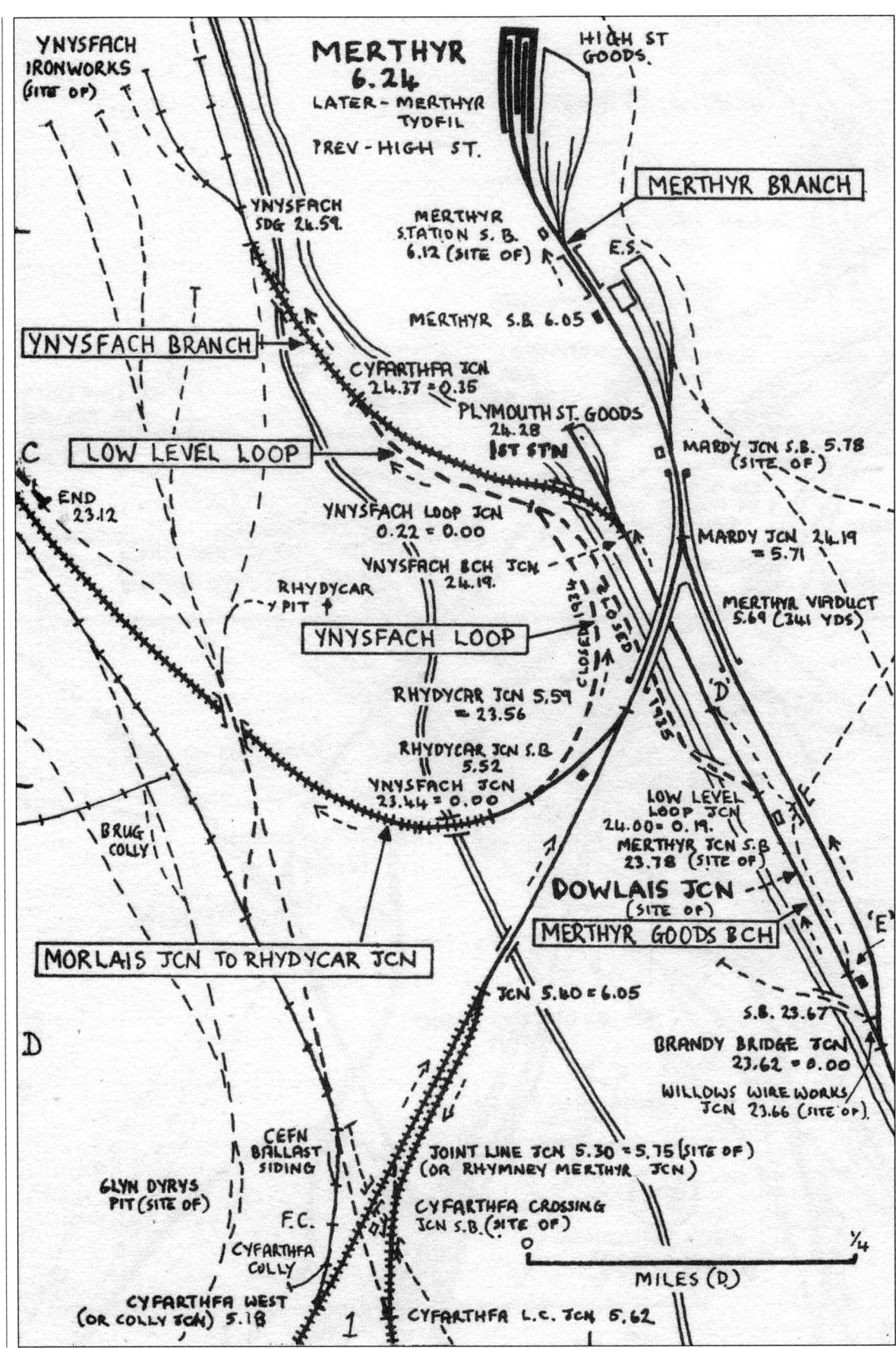

Rhydycar Junction where the B&M and LNW Joint line left the GWR line to Neath with B&M 17 collecting the Webb/Thompson train staff for the section to Morlais Jct. with the 7.50pm Merthyr to Brecon service c.1922. (LCGB/NRM)

A view from Mardy Junction along the former Vale of Neath line to Rhydycar Junction in the distance. (Transport Treasury)

The signals of Rhydycar Junction stand proud as B&M 9 rounds the curve from the Swansea line with a Merthyr to Brecon train which is shown as burning patent Fuel during the Miners' Strike of 1921. Rhydycar Junction SB can be seen in the distance on the left. (LCGB/NRM)

Cyfarthfa Iron Works

Cyfarthfa Ironworks were begun in 1765 by London banker Anthony Bacon, a good 100 years before the B&M Railway reached Merthyr, being dependent on pack horse, then canal modes of transport for the inwards and outwards traffic for the works. A fellow native of Whitehaven, Cumberland, William Brownrigg obtained a lease to mine 4,000 acres of land on the west side of the River Taff at Merthyr for iron ore, coal and limestone and a relation Charles Wood built a forge there to use the potting and stamping process for which he had a patent. Construction of their first coke powered blast furnace began in August 1766, 50 feet high with cast iron blowing cylinders replacing the traditional bellows. This was brought into blast in the autumn of 1767, by which time the Plymouth ironworks had been leased to provide pig iron for the forge.

On the retirement of Brownrigg in 1777, Richard Crawshay became Bacon's partner with responsibility for supplying cannon to the War Deptartment or Board of Ordnance, transferring their manufacture to the ironworks, accompanied by a request for ships to carry the produce from Penarth. The Cyfarthfa Canal, a short waterway nearby, was constructed during the late 1770s to bring local coal to the ironworks and doubtless to help transport produce away, especially after the opening of the Glamorganshire Canal.

In 1782, Bacon, by now a Member of Parliament, had to give up government contracts so passed the forge and boring mill together with its cannon building business to Francis Homfray, but he gave it up in 1784 to David Tanner, so that his sons could found the Penydarren Ironworks. Bacon died in 1786 when Tanner also gave up interest in the works. Bacon's interests were subject to Chancery proceedings which directed that the works should pass to Richard Crawshay. He took out a licence from Henry Cort for the use of his puddling process and proceeded to build

B&M 18 approaching Cyfarthfa with the 2.50pm Merthyr to Newport via Pontsticill as referred to previously c.1922. (LCGB/NRM)

the necessary rolling mill, the problems in so doing being resolved in 1791 when production got underway. In 1790, Crawshay terminated the barely profitable partnership, and managed the works on his own, adding further furnaces in the following years. It was under Richard Crawshay that Cyfarthfa rapidly became an important producer of iron products such as cannon and other weapons of war due to the several naval conflicts the country was engaged in at this time, so much so that Admiral Nelson paid a personal visit to the works in 1802. Indeed, the Crawshay family crest featured a pile of cannonballs in recognition of this important product of the works.

William Crawshay inherited the works following his father's death in 1810 but was less committed to Cyfarthfa than his father, though by 1819 the works had grown to six blast furnaces with an annual production of 23,000 tons. By now, the Industrial Revolution was well under way producing a huge demand for quality iron products, with the Tsar of Russia sending a representative to inspect the production of iron rails. However, during this period, the works was overtaken in importance and production by its long-term rival at Dowlais. The Cyfarthfa Canal, which had been used for the passage of inwards and outwards traffic to and from the works, ceased operation during the late 1830s. William Crawshay also built a grand home which became known as Cyfarthfa Castle (though not recognised in the GWR Castle Class of engines). Erected in 1824 at a cost of £30,000, it featured Norman and Gothic designed turrets and towers and stood in 158 acres of parkland overlooking the ironworks across the river.

There was a considerable amount of freight on the B&M in connection with the iron works at Merthyr, such as this goods passing Cyfarthfa, closed in 1919, hauled actually by B&M 1, though it appears to say No. 11 on the buffer beam – No. 11 was a 2-4-0T. The train is seen crossing the twin-arched tramway bridge at Cyfarthfa en route to Cefn Coed, and is comprised of five plank open wagons of the period, the first two being Midland and the next GW. The sixth vehicle appears to be a horsebox, presumably unoccupied. These normally travelled attached to passenger trains when in use, and is probably returning or going forward empty. (LCGB/NRM)

The last of the great Crawshay ironmasters was Robert Thompson Crawshay who had a short life from 1847-79. Foreign competition and the rising cost of iron ore which now had to be imported through Cardiff Docks and be conveyed uphill to Merthyr, due to the exhaustion of local supplies, had a radical effect on Cyfarthfa works production as did the development of the Bessemer process for producing steel. Robert Crawshay was reluctant to switch the works to the production of steel, unlike others in the area and the works was forced to close in 1875. However, his sons reopened the works after his death in 1879 and converted it to steel production, involving a long closure period until 1884. The brothers continued in business at Cyfarthfa until 1902 but in that year the works was sold to Guest Keen & Nettlefolds, the owners of the Dowlais Plant. With steel production now worldwide, Cyfarthfa was forced to close again by 1910, and though briefly re-opened in 1915 to aid the production of war materials for the 1914-18 conflict, it finally closed again in 1919, then falling into disrepair until dismantled in 1928. The closure of the works was a devasting blow to the local community, many of whom relied on it for their livelihood.

The B&M railway line ran alongside the works with various access points and can be seen as a background to many of the photographs taken by Angus Lewis around the time of the Railway Grouping in 1922-3. The railway fraternity owes a huge debt of gratitude to Angus for recording so brilliantly the early 1920s railway scene in the area, not only recording the Grouping and its effect on the companies operating their railways around Merthyr, but producing valuable views of Cyfarthfa Works in its final state. These views can be seen in *Six Railways to Merthyr*, which is dedicated to the photographs of Angus Lewis in that period. His photographs were later included in the Ken Nunn collection of South Wales and then taken over as part of the Locomotive Club of Great Britain collection, now residing at the National Railway Museum.

B&M 9 with the 11.50am Merthyr to Brecon passes Cyfarthfa ironworks c.1922. The GWR withdrew the 5 former B&M 2-4-0Ts in 1922-24, though they were based on the GW Metro tank design. (LCGB/NRM)

Cyfarthfa Works while still in production. (Merthyr Library)

Cyfarthfa Works was so important to the history of the town of Merthyr Tydfil that several portions of the works have been preserved, including six of the original blast furnaces, the largest and most complete surviving examples that exist of that era, with further excavations revealing other aspects of the working of the plant. The site now forms part of the Cyfarthfa Heritage Area with plans by the Merthyr Tydfil County Borough Council for restoration of the ironworks site.

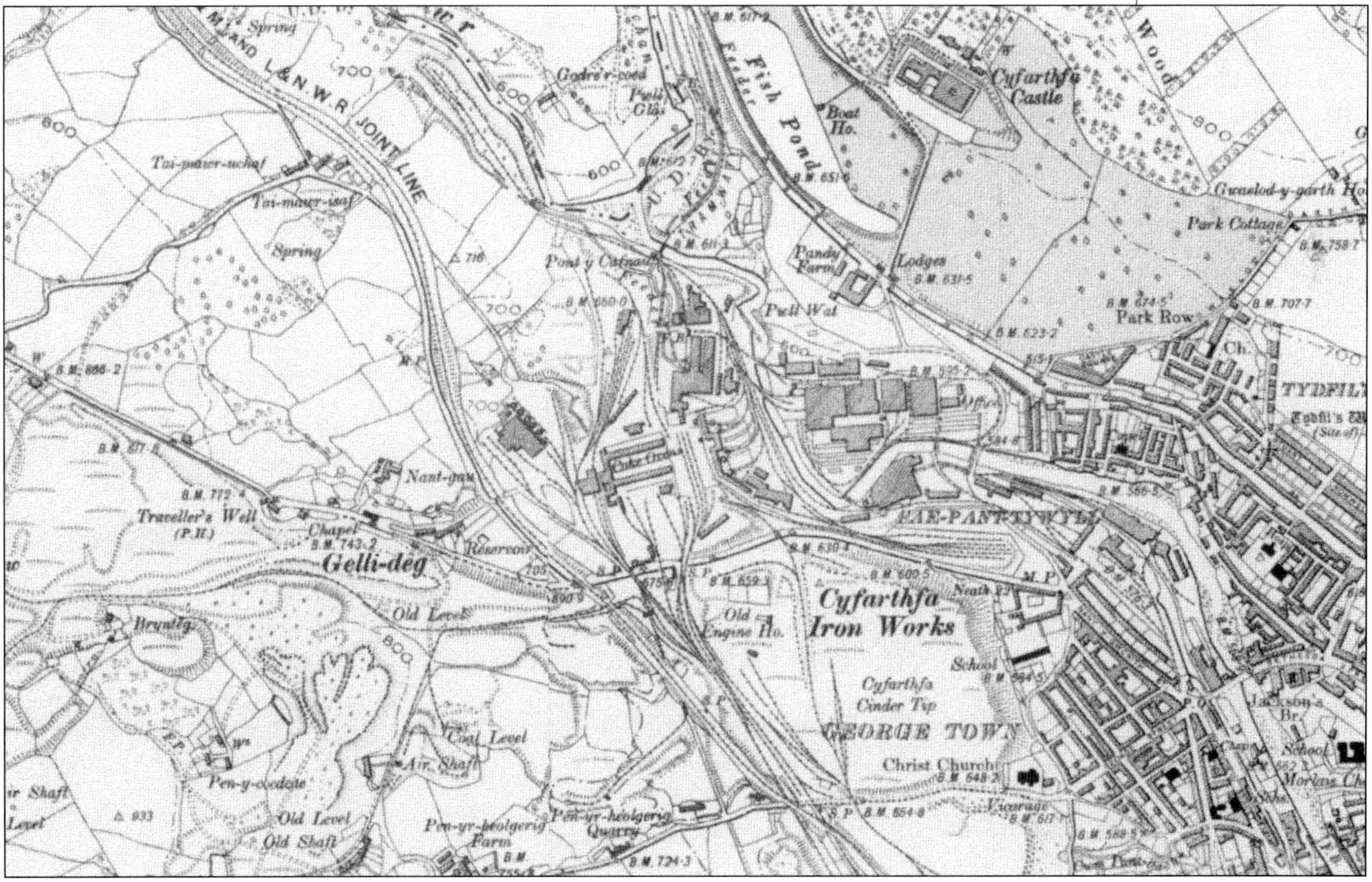

Map of the area around Cyfarthfa Works at the start of the twentieth century. (National Library of Scotland)

Ynysfach Ironworks

The other Ironworks with a close working relationship to the B&M was at Ynysfach where there was a direct railway connection at Rhydycar Junction. The ironworks was opened by Richard Crawshay, who had taken over the Cyfarthfa Works on the death of Anthony Bacon in 1786. There are however reports that a foundry existed on the site in earlier years. It was Crawshay's intention to produce a better-quality refined iron at Ynysfach for processing at Cyfarthfa into a final product, Ynysfach being to all intents and purposes an extension of Cyfarthfa, the pig iron moved between the two works first by tramway and canal and later by railway. The improved quality of iron was achieved by the use of steam blowing engines, the furnaces at Cyfarthfa still being reliant on water power for the blast. The increased demand for this better-quality iron produced a need for expansion and development at Ynysfach and between 1836-39, William Crawshay, who had taken over the works from Robert Thompson Crawshay, carried out significant rebuilding of the plant, including the rebuilding of the northern engine house, provision of two additional blast furnaces, two casting houses, a new refinery and engine house and further tall chimney. A further improvement to the refinery was completed around 1850, all to benefit the quality of the iron supplied to Cyfarthfa. By 1860, the Ynysfach works was at its most productive. Though the furnaces were reconditioned in 1884, the conversion of the Cyfarthfa plant to steel production rendered the Ynysfach iron production unnecessary and the plant was held in reserve until demolition began in the first decade of the new century. Various sections of the works have been restored as the Ynysfach Iron Heritage Centre.

Heolgerrig Halt

Heolgerrig Halt was south of Llwyncelyn Jct. and north of Ynysfach Jct. and Cyfarthfa Jct. where it joined the Ironworks lines. It opened as a halt on 31 May 1937 and closed on 13 November 1961.

Heolgerrig Halt. The Halt was opened by the GWR in May 1937 and closed on the withdrawal of the passenger service in November 1961. (Michael Hale/GWT)

LOCATION ANALYSIS • 29

Heolgerrig Halt seen from the footplate of 4555 on 3 May 1964. (P.J. Garland/Roger Carpenter)

Llwyncelyn Junction

At 22m 70ch from Brecon this was the junction between the B&M and the Cyfarthfa and Ynysfach Ironworks and their associated private railways. As these fell out of use, the box became obsolete and was closed in August 1927.

Llwyncelyn Junction SB and road bridge to Colliers Row in the 1920s. Connections here fell out of use with the closure of the Iron Works and ancillaries, leading to the SB closing in August 1927. (Ken Mumford Presentations)

Heading past Llwyncelyn, a Brecon to Merthyr service with 3595 in charge. (LCGB/NRM)

The green fields of Llwyncelyn are the setting for this shot of an interesting formation for the 2.5pm Brecon to Merthyr worked by B&M 18, probably the return working of the 2.50pm Merthyr to Pontsticill with three vans on the front of the four coach train, the first vehicle being a GW Horsebox, followed by what is believed to be a B&M Market Van, then the normal full brake. Date c.1922. (LCGB/NRM)

Approaching Llwyncelyn, 2113 on a Merthyr to Brecon working, which was the first working of a B&M train by a GW engine after the Grouping. (LCGB/NRM)

The Barry to Llandrindod Wells service here has a 2301 for power beyond Merthyr after the Grouping as it approaches Llwyncelyn, and could now work through to destination. (LCGB/NRM)

GW 2021 Class 2113 has the 4pm Merthyr to Brecon approaching Llwyncelyn Junction after the Grouping in 1923. (LCGB/NRM)

Cefn Coed

At 21miles 29ch from Brecon, this was the main station and crossing point on the line. Known as 'Cefn' until May 1920 and remembered chiefly for its excellent viaduct at the south end and its large informative nameboard exhorting passengers to change between B&M and LNW trains there. Goods facilities were available at the south end of the Up platform and were withdrawn on 4 May 1964, when concentrated at Merthyr, and the sidings disconnected soon afterwards.

Map of the Cefn Coed area at the start of the twentieth century. (National Library of Scotland)

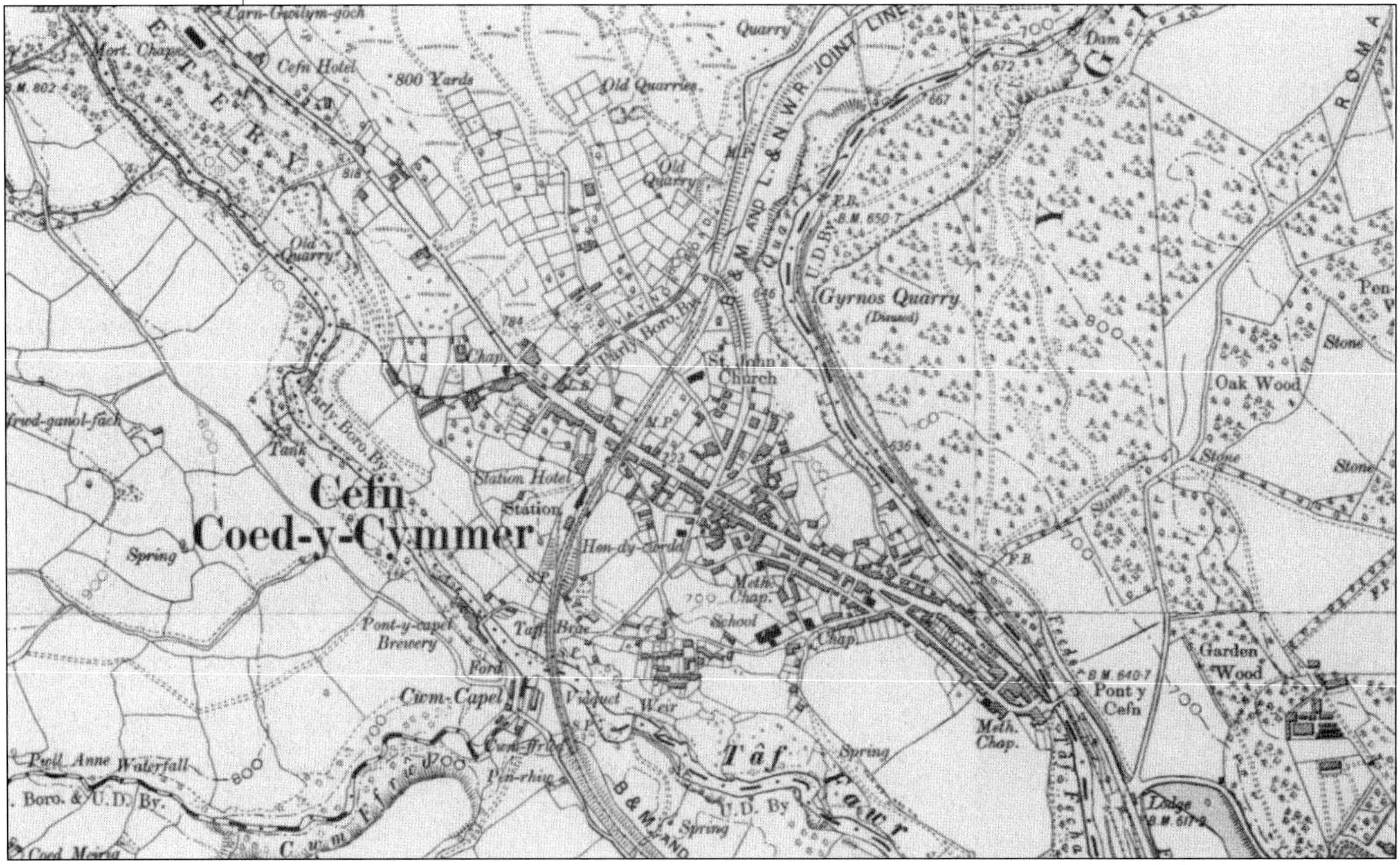

LOCATION ANALYSIS • 33

Cefn Coed station as seen from last day special on 3 May 1964. (P.J. Garland/Roger Carpenter)

Cefn Coed station as on 13 September 1951. (H.C. Casserley)

Merthyr to Pontsticill auto train arriving at Cefn Coed. (Ken Mumford Presentations)

Another view of Cefn Coed, this time from the elevated station footbridge as 6433 works the one coach 10.02am Pontsticill to Merthyr auto service on 4 November 1961. (W.G. Sumner)

LOCATION ANALYSIS • 35

The ornate Cefn Coed station nameboard. (Ken Mumford Presentations)

Cefn Coed station as at 4 August 1963 after the closure of the station to goods traffic. (Garth Tilt)

B&M 9 enters the station loop for Cefn Coed with what is claimed to be the 6.30pm from Brecon to Merthyr c.1922. (LCGB/NRM)

Having just crossed the Viaduct, B&M 18 enters Cefn Coed station with the 2.50pm Merthyr to Newport which will join up at Pontsticill with a Brecon to Newport service. c.1922. (LCGB/NRM)

A Merthyr to Brecon service departing Cefn Coed for Brecon c 1922 with GWR 3595 in charge. (LCGB/NRM)

LOCATION ANALYSIS • 37

Two views of the morning Merthyr to Pontsticill Goods at Cefn Coed, passing the 10.02 Pontsticill to Merthyr in the other platform on a wet morning in November 1961. 9675 is working the Goods with 9618 the passenger. (W.G. Sumner)

Cefn Coed in the winter sun on 4 November 1961 as 6433 works the 10.02am Pontsticill Jct. to Merthyr, with more staff than passengers evident. (W.G. Sumner)

A view of Cefn Coed Viaduct looking towards Merthyr on 13 September 1951. (H.C. Casserley)

The layout at the Cefn Coed station end of the viaduct as on 3 May 1964. (P.J. Garland/Roger Carpenter)

Above left: Cefn Coed SB on 3 May 1964, the last day of service, other than recovery trains. (P.J. Garland/Roger Carpenter)

Right: Cefn Coed water column on 3 May 1964. (P.J. Garland/Roger Carpenter)

Vaynor Siding

This siding, at 20miles 21ch from Brecon, served Vaynor Quarry and appeared as such in B&M timetables. In 1895 a stone train is shown as working to Llwyncelyn Jct. and return, presumably conveying limestone for Cyfarthfa Works but is absent from the supplement for 1902. It was presumably then served on demand as the siding apparently did not close until August 1966, the line from Merthyr by then worked as a siding.

Pontsarn

At 19miles 64ch from Brecon, Pontsarn, a single line platform with a chequered history, having been a request stop from 1869, then later a staffed station, was renamed Pontsarn for Vaynor and was finally downgraded to a halt by the GWR in 1934. Between the Grouping and the 1939 War, it was used as a recreational venue for schools, churches etc. on a day-out basis with a large field behind the station used for that purpose.

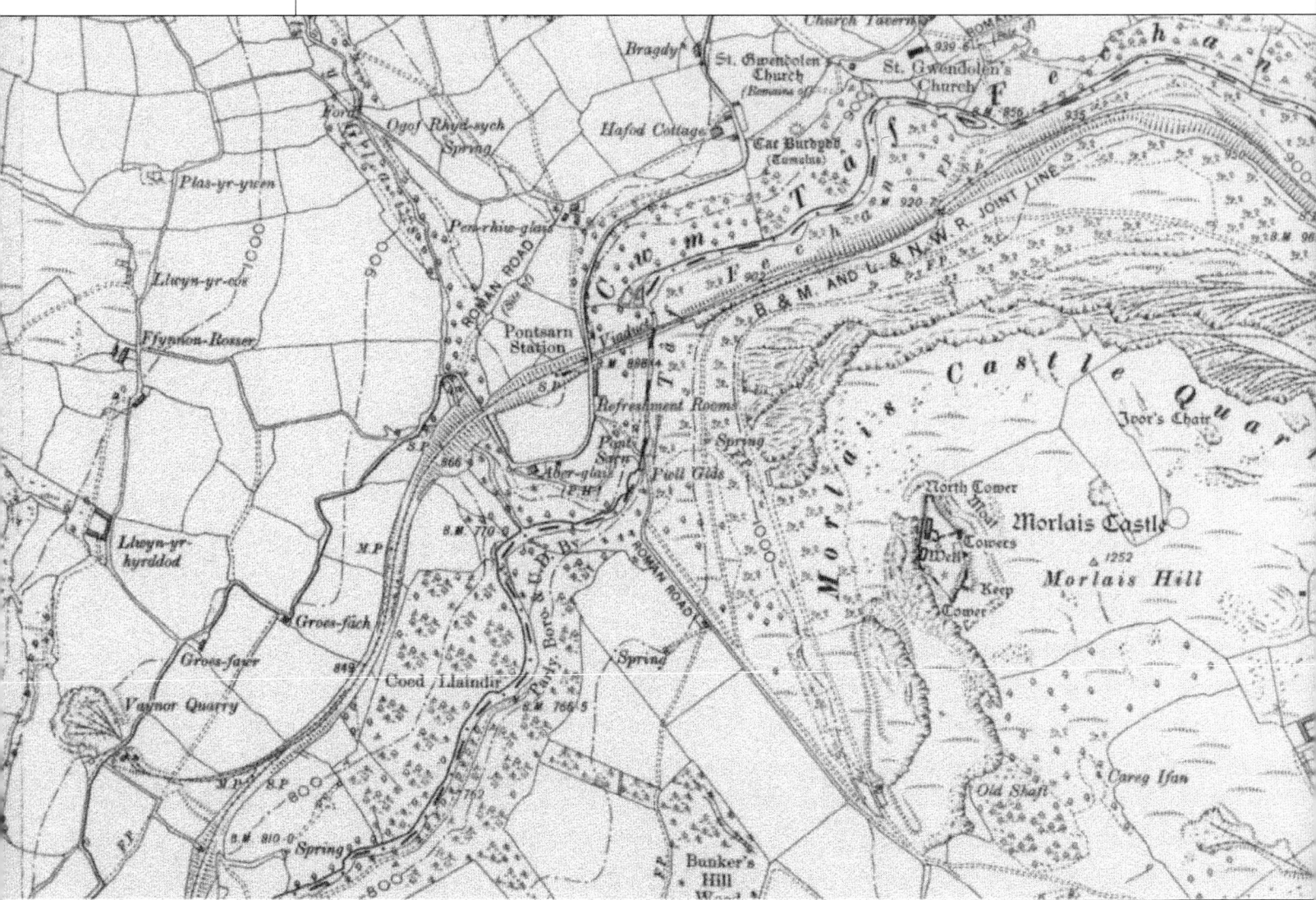

The Pontsarn area at the start of the twentieth century. (National Library of Scotland)

LOCATION ANALYSIS • 41

Pontsarn station platform after closure, seen on 11 April 1963. (Jeff Stone)

Pontsarn Station with the Station Master and staff posing alongside a passenger and child. Note the number of enamelled signs on view. (Lens of Sutton)

Pontsarn platform with the 10.02am Pontsticill Jct. to Merthyr in auto mode with Merthyr's 6433 on 4 November 1961. W.G. (W.G. Sumner)

A wider view of Pontsarn platform with 6433 on the 10.02am Pontsticill Jct. to Merthyr service on 4 November 1961. (W.G. Sumner)

Pontsarn station with an auto service to or from Merthyr. In the background is the large building and field that provided facilities for the many summer visitors. (Ken Mumford Presentations)

Pontsarn Viaduct seen from the 10.02am Pontsticill Jct. to Merthyr on 4 November 1961. (W.G. Sumner)

Pontsarn Viaduct shown in full length on 11 April 1963, and a section of it shown artistically on 7 July 1962. (Jeff Stone, Alan Jarvis/SLS)

After the Grouping this former Taff Vale M1 482 found itself working from Merthyr to Brecon, here seen crossing Pontsarn Viaduct c.1923. (LCGB/NRM)

Pontsarn Viaduct with the 12.10pm Merthyr to Brecon crossing around the Grouping. (LCGB/NRM)

The final SLS Special on 2 May 1964 passing the long closed Pontsarn station, seen from the overbridge at the north end of the platform. (John Hodge)

Morlais Junction

This Junction was 18miles 75ch from Brecon and was the point where the B&M joined the double LNWR track from Abergavenny. The latter had suffered much the same problems of access to Merthyr as the B&M, eventually coming to an agreement to jointly use the B&M route. Formal agreement was reached in 1875 and LNWR services to Merthyr began in June 1879. Following the Grouping, the GWR renamed the location Morlais Tunnel Junction to avoid confusion with Morlais Junction near Llanelly.

The point at which the B&M line met the LNW coming north through Morlais Tunnel before sweeping south jointly with the B&M through Pontsarn, and Cefn Coed into Merthyr, joining up with the GWR (former VoN) line from Neath at Rhydycar junction. The LNW MT&A line proceeded through the tunnel to Pantysgallog LL and then on to Dowlais High Street, under the B&M main line near Dowlais Top and on to Rhymney Bridge. The B&M train is the return of the 2.50pm Merthyr to Pontsticill c.1922. (LCGB/NRM)

Morlais Junction SB where the LNW line to Abergavenny splits from the B&M line to Pontsticill, as the signalman exchanges the staff with the fireman of a Merthyr to Pontsticill train on 22 August 1957.
(Ian L. Wright)

Morlais Junction, showing the signal box, junction of the B&M and MT&A routes and Morlais Tunnel south portal, a later photograph, showing Great western signals.
(Ken Mumford Presentations)

An LNWR Coal Tank leaving Morlais Tunnel and heading towards Merthyr with a train of coal empties. (LCGB/NRM)

Pontsticill Junction

The line from Pontsticill to Merthyr opened as far as Cefn Coed on 1 August 1867, and was extended through to Merthyr High Street, the original Vale of Neath station, on 1 August 1868. Between Pontsticill Junction and Pant the main B&M line ran parallel with the Merthyr to Pontsticill Jct. branch, this rising gradually to the level of the main line as it approached the junction. This was also the beginning of the very scenic nature of the line as it now ran alongside the reservoirs at the south end of the Brecon Beacons, making it one of the most attractive stretches of line in the country. Pontsticill Junction station lay at 29miles 63ch from Newport at 1 in 198 facing north and was the main station on the line between Bargoed and Talyllyn Jct., the meeting point for trains on the main Newport to Brecon route and those coming in from Merthyr. Dating from the early years of the B&M when much freight and passenger traffic was exchanged between the two lines at this station, there were several sidings both on the Up and Down side, including a turntable for use when tender engines were involved, largely off the Mid Wales line, or when tank engines required to be turned in connection with the Seven Mile Bank. Through freight trains from Merthyr to Brecon and reverse and the later Bassaleg to Brecon service were always booked to work at Pontsticill to exchange traffic between the two lines, while the later Dowlais Central ammonia tank trains were often recessed there to allow a passenger train to pass. The signal box controlling all movements at this location, which was renewed in 1885, was located at the south end of the Down platform, on which the main facilities were located, including the station master's house. The Up platform was an island, the west face of which was used by the terminating passenger services from Merthyr, the east face being used by the through Up Brecon services and northbound freights.

In the early years of the twentieth century, and probably before, there were quarries in the high ground on the east side of the line, with tramroads running down to loading points alongside the main line. The Abercriban Quarry had two loading points, one just north of the station with a PSA with the B&M in the name of M. Richards from 1901 to 1934. The other, further north, had a PSA for A.W. Lewis trading as the Abercriban Quarry Co. from 1913-17. Robert McAlpine is also recorded as having a PSA between 1915 and 1926, but whether this referred to the above sidings is unclear.

Further south, on the east side of the line to Pant, was the Baltic Quarry, again with a tramroad bringing stone down to a loading point on the main line, with a PSA to the New Tylerbont Stone & Asphalt Co. Ltd. between 1925 and 1944, though probably existing before the turn of the century. A serious landslip occurred near this loading point in November 1931.

On the single line to Merthyr, which fell away to a lower level on the west side of the main line, there was a loop located opposite the loading point on the main line, but this was declared surplus to requirements and was removed in July 1925, having previously been used to hold freight trains awaiting acceptance into Pontsticill.

Further south towards Pant was another quarry Tylahaidd, dating from around 1875 which was closed and re-instated by June 1896, but no further details are available.

A Map of the Pontsticill area around the start of the twentieth century. (National Library of Scotland)

The station nameboard in its attractive location on the south end of the island platform on 27 September 1958. The platforms were always well presented and the flower beds and cleanliness bear witness to the care of the staff.
(R.J. Buckley/Initial Photographics)

A fine shot of Pontsticill Junction station in the sun on 13 October 1962. The normal practice was for the island platform to be used for the Merthyr service with the Brecon services using the up and down main platforms. The GWR rebuilt the SB top and installed a 52 lever frame in 1930. (W.G. Sumner)

LOCATION ANALYSIS • 51

A view of the station in the early years of the century with a saddle tank on a goods service, probably for Merthyr, and showing the wooden passenger footbridge, which, with that at Machen, were the only two on the line. Note the milk churns which were conveyed in quantity at that time. The footbridge is said to have been damaged beyond repair by a steam crane in early GW days and never replaced. The somersault signals and original McKenzie & Holland SB top are worthy of note. (D.K. Jones Collection)

Pontsticill Junction main platforms as seen from the south end of the station with a Merthyr train on the island platform on 3 May 1951. Often the stock provided was an auto coach as here, though not always an auto-fitted 64XX, a 57XX deputising. (R.C. Riley/Transport Treasury)

Ebbw Junction's 2227, with an ROD tender, calls at Pontsticill with a Brecon to Newport service on 11 September 1951. (H.C. Casserley)

The 6.15pm Brecon to Newport approaches Pontsticill on 22 August 1957 behind Ebbw's 3634. The milk tanker shown worked daily from Builth Wells to Aberdare via Talyllyn Jct., Pontsticill Jct. and Merthyr, so will shortly be detached. The van behind it is not identified in the current coach working programme. (Ian L. Wright)

The last year of the Newport-Brecon service was shared between Ebbw Junction 2251 and 8750 Class (later version) engines. Here their 9667 worked a southbound train on 18 August 1962.
(F.K. Davies)

Merthyr's 4690 has shunted its returning freight from Brecon onto the up line, presumably to await its path home on 13 October 1962.
(W.G. Sumner)

A 1940s picture of the staff at the station with a light engine at the up platform and a Merthyr connection waiting at the other side of the island platform. The Station Master is in the centre of the group in his GWR gold-braided cap with the signalman completing the quartet in his box. (Ray Caston Collection)

A pleasing view of the whole station from the south end on 9 June 1962 with Ebbw Jct.'s 2218 heading a southbound service to Newport. (Colour Rail)

LOCATION ANALYSIS • **55**

In this 1948 view of Pontsticill, the 3.15pm to Merthyr (connecting out of the 2.10pm from Brecon) departs from the down main platform to save passengers crossing the line to the island platform. 5711 is working the service with clerestory roof stock in the station sidings on 29 March 1948. (SLS)

A 64XX waits with a connecting service to Merthyr as 7771 runs into the down main platform with a Brecon to Newport service in the early 1950s. (SLS)

The layout accessing the Merthyr branch at Pontsticill as on 8 September 1962. (Michael Roach)

A typical winter scene at Pontsticill looking south at the Merthyr branch (right) and the line to Pant (left), on 27 December 1962. (Jeff Stone)

LOCATION ANALYSIS • 57

A view of the south end of the island platform and south end layout from a Brecon to Newport service on 10 July 1958. In the distance, beyond the junction, there were two further long sidings on the up side of the main line, including a rail weighbridge used in connection with limestone traffic from the several quarries in the area. (H.C. Casserley)

The layout at the south end of the station seen from a departing train on the 18 July 1959 with the empty Ammonia tank train waiting to come into the station behind. (R.E. Toop)

2218 approaches Pontsticill with the 3pm Newport to Brecon on 1 September 1949. The veteran passenger brakevan on the front of the train is the Thursday Only Swindon Stores Van bound for Brecon. (B.W.L. Brooksbank/Initial Photographics)

The SLS Special on 2 May 1964, hauled by privately-owned 4555 and 3690, takes water at the south end of Pontsticill en route to Dowlais Central. (F.K. Davies)

4611 has the 2.5PM Brecon to Newport as it arrives at Pontsticill the same day as previous picture. (John Hodge)

LOCATION ANALYSIS • 59

An early 1950s shot of 7771 working a Brecon to Newport service calling at Pontsticill. (SLS)

A fine view of the return empty Dowlais ICI Ammonia tanks returning double headed by 5793 and an 875X through Pontsticill, having joined the former B&M line at Talyllyn East Junction from Hereford in the early 1950s. (M. Whitehouse Collection)

Setting off south from Pontsticill Ebbw Junction's new acquisition 2240 has the 2.15pm Brecon to Newport. The engine was soon replaced as it was condemned shortly after this working on 2 June 1962. (Alan Jarvis/SLS)

The Merthyr platform on 1 May 1956. (T.C. Cole)

The north end layout of the station on 2 May 1964. By this time, the layout on the right was out of use. (P.J. Garland/Roger Carpenter)

The layout at the north end seen from a 2251 hauled 2.10pm Brecon to Newport service on 17 September 1962, as a connecting service to Merthyr runs into the other platform. (R.E. Toop)

Standard 2MT 46522 leaves Pontsticill with the 11.15am Newport to Brecon early in 1962. The stone-built B&M water column is prominent at the end of the platform. (John Hodge)

The Brecon to Bassaleg Goods, composed mostly of empty minerals off domestic coal traffic, leaves the single line to run through Pontsticill Jucntion on 12 October 1962… and is also seen at the south end of the station, with Ebbw's 2298 in charge. (W.G. Sumner)

LOCATION ANALYSIS • 63

...and awaits the signal to proceed south. (Colour Rail)

This Brecon to Merthyr Goods service, mostly composed of empty vans and hauled by 8776, runs through the station probably in the late 1950s. 8776 (without a shedplate) was one of Canton's quite large contingent of panniers and was probably on loan to Merthyr to cover a shortage there, no record existing of a transfer, the engine being withdrawn from Canton in December 1962. (SLS)

64 • RAILWAYS & INDUSTRY ON THE BRECON & MERTHYR: MERTHYR-PONTSTICILL JUNCTION-BRECON

Entering Pontsticill from Dolygaer, Brecon's 2236 at the head of the Brecon to Bassaleg Goods in the early 1960s. (Colour Rail)

9644 takes water on the 2.5pm Brecon to Newport as a two coach train from Merthyr stands in the up main platform on 19 June 1962. (Garth Tilt)

Running alongside the reservoir departing Pontsticill Jct. 9644 heads the 3 coach 8.3am Newport to Brecon on 4 November 1961. The line in the foreground ran to the north quarry, worked by the Abercriban Quarry Co. (Malcolm James)

The signalbox and former Station Master's house are still standing but little else in this view of the the site of Pontsticill station on 15 October 1967, before the arrival of the Brecon Mountain Railway. (Garth Tilt)

66 • RAILWAYS & INDUSTRY ON THE BRECON & MERTHYR: MERTHYR-PONTSTICILL JUNCTION-BRECON

LOCATION ANALYSIS • 67

Opposite and left: Three views of the platforms and layout at Pontsticill station with a train from Merthyr at the main up platform on 8 September 1962.
(Michael Roach)

The empty Dowlais ICI tanks which had been recessed at Ponsticill restart the final stage of their journey with 4635 on 2 June 1962.
(Alan Jarvis/SLS)

Surrounded by the lovely Pontsticill scenery, the Merthyr auto service awaits departure. (Colour Rail)

In conventional mode, Merthyr's 9643 has run round its single coach train composed of W102 and awaits departure at the down main platform with the 1.19pm to Merthyr on 24 March 1951. (W.A. Camwell/SLS)

LOCATION ANALYSIS • 69

The track on the Merthyr branch as it approaches Pontsticill Junction. The running line from Merthyr is on the left as it rises towards the junction. The disused stone sidings are in the foreground with the south end of the reservoir in the background of this 1964 shot. (Jeff Stone)

With the reservoir in the background 3661 leaves Pontsticill for Newport. (Alan Jarvis/SLS)

The SLS Special of 2 May 1964 taken the Merthyr line on the final stage of its tour from Merthyr to Brecon and back. (Alan Jarvis/SLS)

Approaching Pontsticill on the single line from Pant 3706 heads a Newport to Brecon service in the early 1960s. (Colour Rail)

Approaching Pontsticill Merthyr's 9675 heads a one coach service from Merthyr on 20 April 1960. (T.B. OWEN/Colour Rail)

LOCATION ANALYSIS • 71

A Newport to Brecon service with 9674 with a Merthyr to Pontsticll connection with 6416 on 30 September 1961. (Mark Warburton)

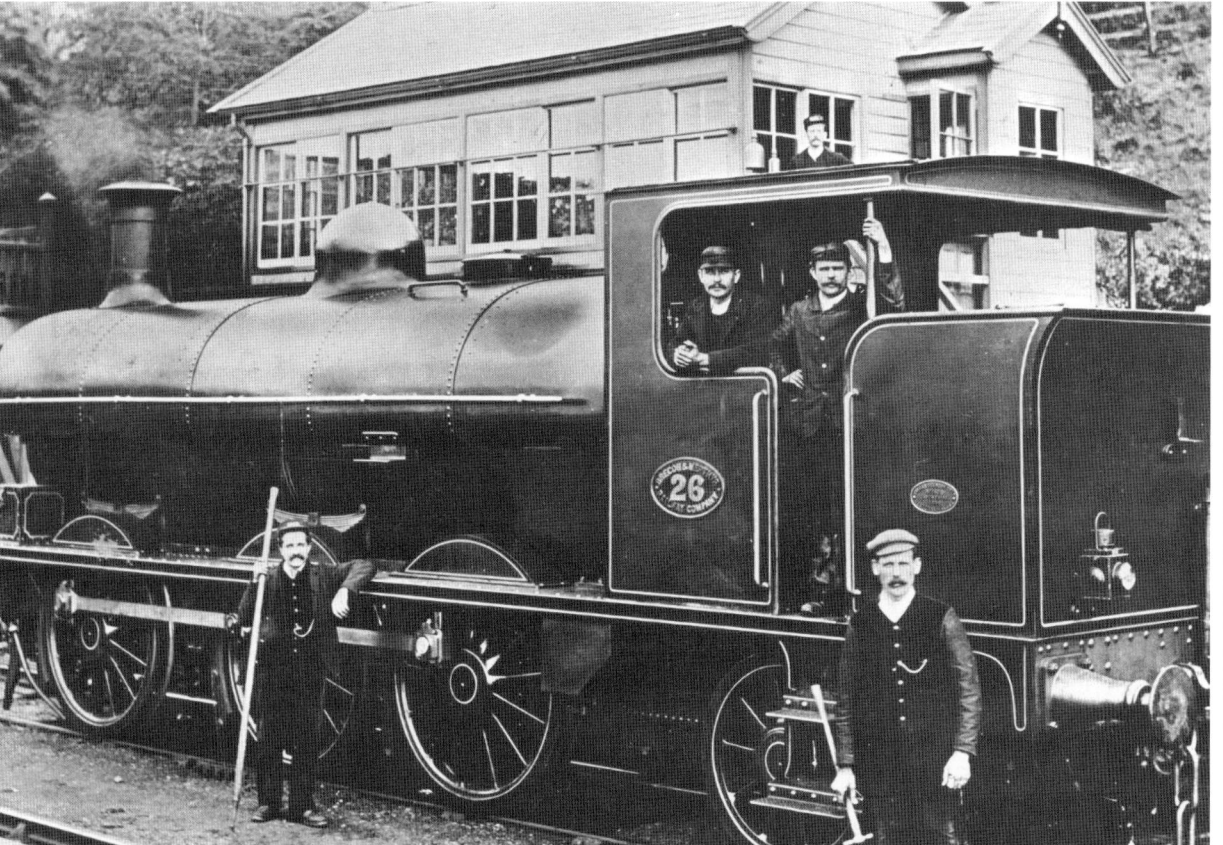

To finish our trip to Pontsticill, a blast from the past. A lovely view of B&M 26 of 1896 vintage in a posed picture with its crew at the south end of the station. (Ray Caston Collection)

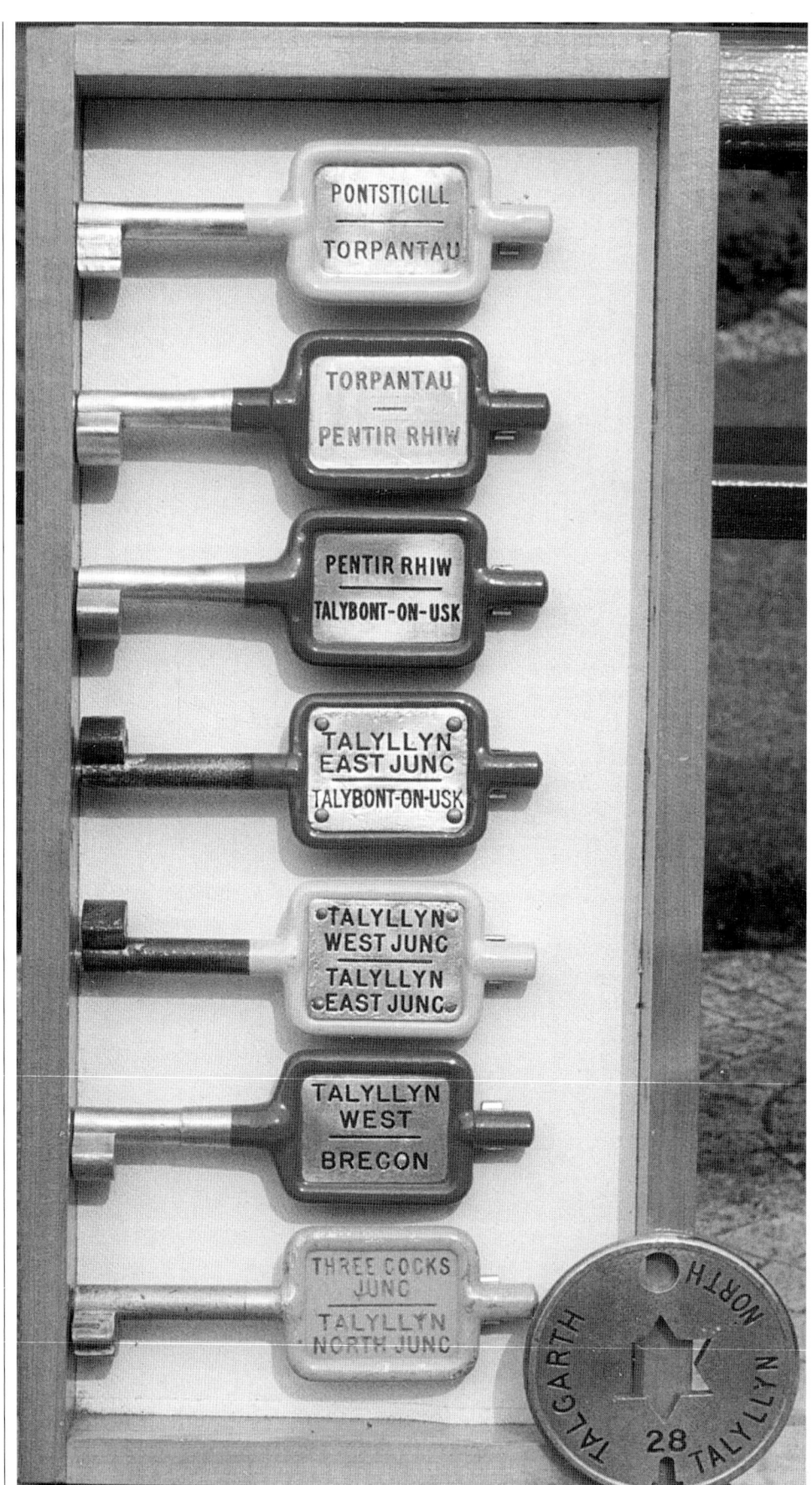

Key Tokens for the single line sections between Pontsticill and Brecon.
(Private Collection)

Dolygaer

Standing at 31miles 36ch from Newport, this was a single-platform station with a small cement-rendered building. It is not clear whether it was staffed in earlier days, other than on event days. In later days it is described as a halt. Its moments of glory date from the earliest days of the B&M when regattas were held on the nearby Pentwyn reservoir. It was common for several thousand people to be conveyed to Dolygaer from the Merthyr and Dowlais areas for these events, staff from Brecon being drafted in to deal with the throng.

An 1885 plan shows a single siding, facing towards Pontsticill, which in 1922 was supplanted by loading banks adjoining the running line, whilst Pentwyn reservoir was enlarged into the Taf Fechan reservoir, opened in 1927. During the heady days of the Pentwyn regattas, there was a hint of a SB being provided, but this did not come to fruition.

The station is now in private hands with the narrow-gauge track of the Brecon Mountain Railway running past on the formation of the former B&M.

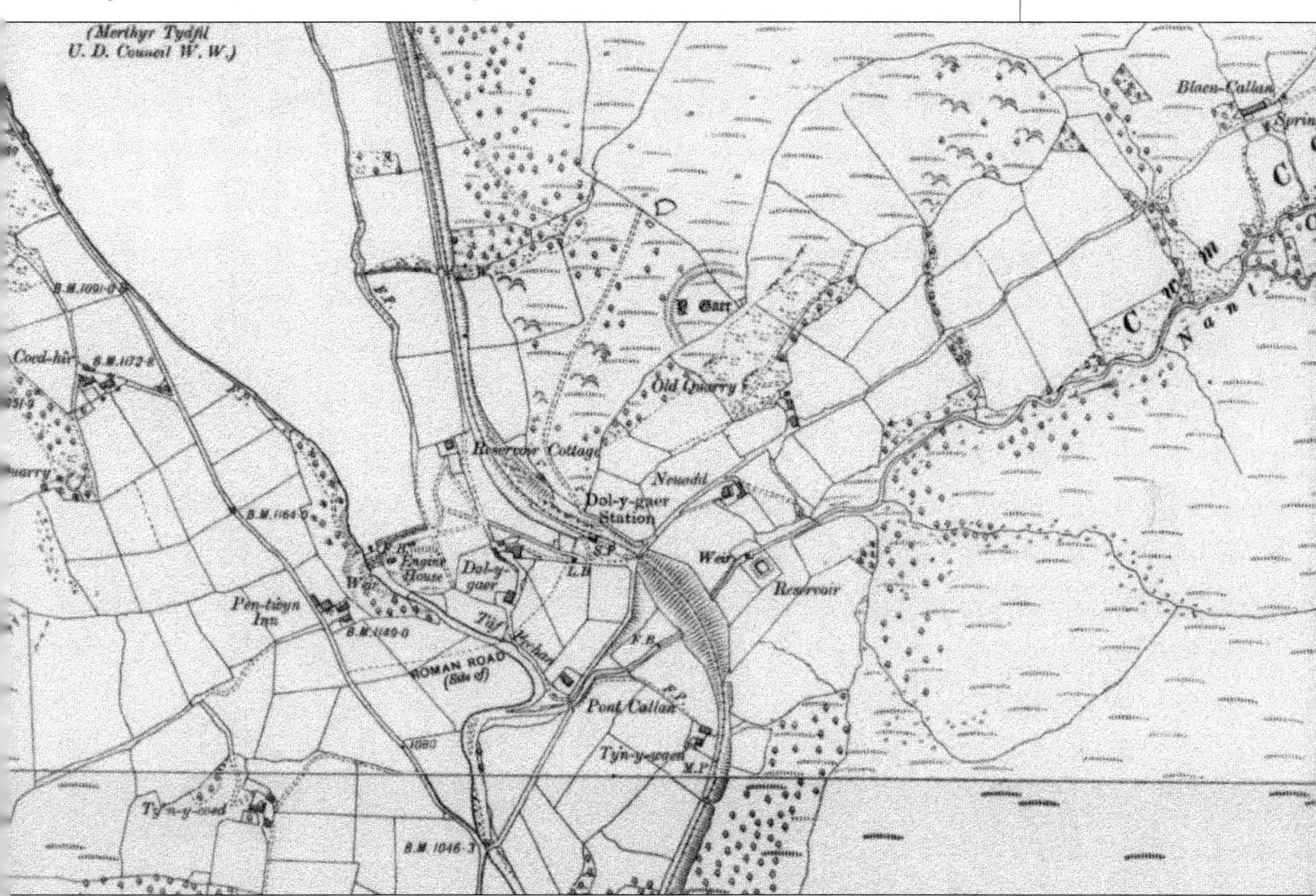

A map of the Dolygaer area at the start of the twentieth century. (National Library of Scotland)

Dolygaer looking south… and looking north. (Lens of Sutton)

The newly acquired 3201 was much used on the Brecon service and is seen here at Dolygaer with a Newport to Brecon service in 1962. (Ken Mumford Presentations)

A view along the platform on 13 July 1962 with the 11.15am Newport to Brecon calling.
(Robert Darlaston)

Dolygaer looking south towards Pontsticill on 8 September 1962.
(Michael Roach)

Dolygaer looking north towards Torpantau on 8 September 1962. The ganger's hut and trolley are prominent in the foreground. (Michael Roach)

The 2.5pm Brecon to Newport with 2298 calls at Dolygaer on 29 September 1962. (W.G. Sumner)

2280 departs south from Dolygaer with the 2.15pm train from Brecon to Newport on 16 May 1959.

The scenery south of Dolygaer is some of the best on the line. Here 2247 leaves the station with the 2.5pm Brecon to Newport on 13 October 1962.
(W.G. Sumner)

78 • RAILWAYS & INDUSTRY ON THE BRECON & MERTHYR: MERTHYR-PONTSTICILL JUNCTION-BRECON

The same train, with passengers leaning out of the windows to enjoy the view, departs with 46522 on 18 July 1959. These engines worked the Brecon turns until they were covered from Ebbw Junction. (R.E. Toop)

The 2.5pm Brecon to Newport is seen in the distance between Dolygaer and Pontsticill with 2247 in charge on 13 October 1962. (W.G. Sumner)

Pentwyn Reservoir at Dolygaer, a considerable attraction for rail excursions in the Victorian era. (Ken Mumford Presentations)

4679 heading into Dolygaer from Pontsticill with a Newport to Brecon service on 8 September 1962. (Michael Roach)

A section of the Seven Mile Bank, once the bane of engines and crews, now a serene section of the Taff Trail which runs all the way from Brecon to the outskirts of Cardiff. (Ken Mumford Presentations)

Torpantau

Located at 33m 2ch from Newport, a crossing loop was provided here for the operating convenience of the railway. Almost from the start of passenger services, passengers alighted here to walk in the mountain scenery, so platforms were provided in response to need. The station gained a certain notoriety when it became known that the Traffic Manager had petitioned the directors for a fireplace to be provided in the SB in winter, this at an altitude of some 1,200 feet.

Torpantau Tunnel (also known as Summit or Beacon Tunnel) was immediately north of the station, taking the line round in an easterly direction into Glyn Collwyn and was described as the highest railway tunnel in the country. More modern signalling was provided in 1892 with the construction of a properly interlocked SB, necessitating the shortening of the passing loop to meet statutory requirements. The two Up sidings were probably provided at this time. In 1918, a tramway was constructed for loading timber, though only lasting a few years and probably in response to wartime needs. The sidings were taken out of use in June 1960, the passing loop and SB following in January 1963, after cessation of passenger operations the previous month.

Today the Brecon Mountain Railway terminates here, though on a different alignment to provide level platforms, the previous station area having been altered by earth-moving to become unrecognisable.

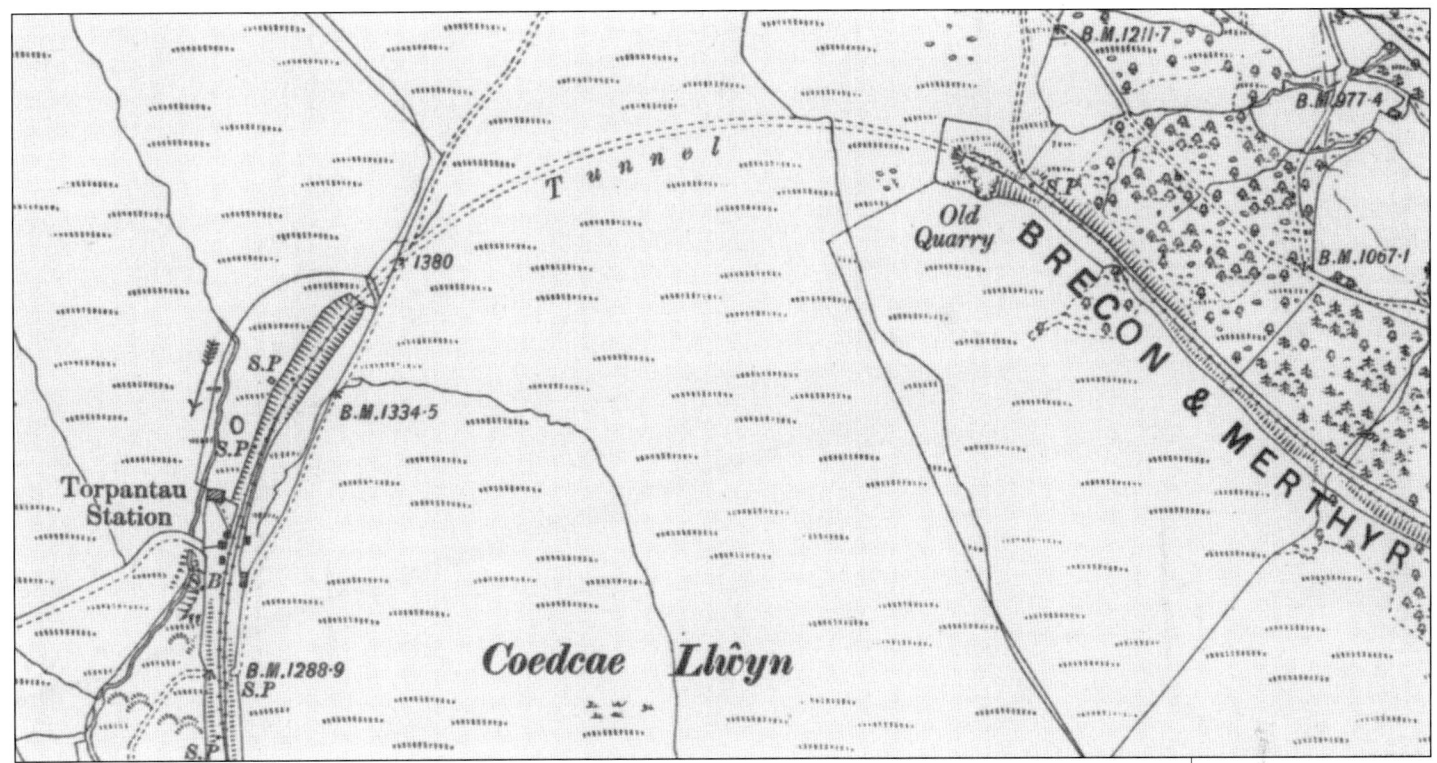

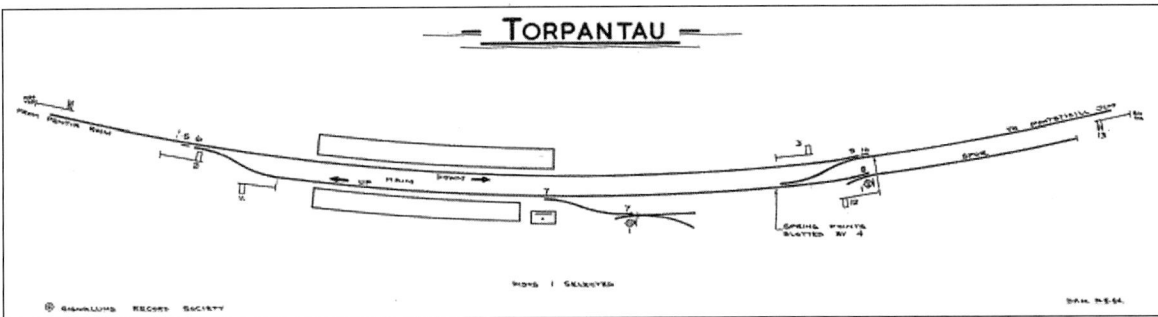

Top: OS Map of the Torpantau area at the turn of the twentieth century., showing the Summit Tunnel and start of the Severn Mile Bank (right). The overall trajectory of the line is noteworthy as it turns in a half moon shape ending at Torpantau station.
(National Library of Scotland)

Above left: Torpantau Track and Signal Layout.

Torpantau station in August 1961.
(Lens of Sutton)

The station nameboard on 26 March 1951. (Ian L. Wright)

A typical winter scene at Torpantau, as in December 1962. (John Hodge)

Fast backwards to the same view from a Brecon bound train in summer on 18 July 1959. (R.E. Toop)

Two views of the 12.10pm Brecon to Newport with Ebbw Junction's 8766 in charge, first arriving at Torpantau and then in a scenic view of the station on 18 May 1959. By then the signal box and loop had been taken out of use.
(R.O. Tuck/A. Jarvis)

8766 on the 12.10pm Brecon to Newport crosses with the 11.15am Newport to Brecon at Torpantau in these two shots on 18 May 1959. (R.O. Tuck)

LOCATION ANALYSIS • 85

The view northwards from half way along each platform on 1at May 1956. (T.C. Cole)

3770 enters the station with a Brecon to Newport service. (W.A. Camwell/SLS)

2298 has the 2.5pm Brecon to Newport seen standing at Torpantau on 29 September 1962. (W.G. Sumner)

3712 arriving at Torpantau with the 12.15pm Brecon to Newport, passing the stone-built water tank. This was the return working of the 8.3am Newport to Brecon on 13 July 1957. (SLS)

LOCATION ANALYSIS • **87**

2227 has just passed through the station and now rejoins the single line on to Pentir Rhiw with the Bassaleg to Brecon Goods on 1 May 1956. (T.C. Cole)

2280 on an up train, passing the down starting signal by the Torpantau road overbridge.

3661 stands at Torpantau with the 12.10pm Brecon to Newport on 16 May 1962. (A.Cooke/Colour Rail)

Below left: Torpantau SB in the early 1900s. (Ray Caston Collection)

Below right: A Merthyr to Brecon Goods passes the 7.35am Brecon to Newport Passenge, both with modern Pannier haulage, the Goods service standing in the station on 2 July 1960. (D.K. Jones Collection)

In the years up to 1962, the Brecon to Newport working was shared between those two depots, but the Brecon engines were often provided by Oswestry in some timetables, probably to help build up the mileage. This could have involved Moat Lane, a sub-shed of Oswestry, as part of a 3 day diagram working Goods from Moat Lane to Talyllyn on Day 1, Brecon to Newport passenger on Day 2 and goods back from Talyllyn to Moat Lane on Day 3. Here their Dean Goods 2409 prepares to start away from Torpantau with the 11.15am from Newport on 14 May 1949. (D.K. Jones Collection)

A rear view of the 2251 Class hauled Bassaleg to Brecon Goods rejoining the single line north of Torpantau on 13 June 1959. (Alan Jarvis/SLS)

The return Dowlais Ammonia tank empties run through Torpantau station with Hereford's 8781 and a 37XX in charge with both crews eager to be in the photograph. (W.A. Camwell/SLS)

Freight trains cross as a returning pair of panniers with the Brecon to Merthyr freight composed mostly of empty mineral wagons, passes an up freight service hauled by a single pannier and conveying coal as well as van traffic on 14 July 1961. The leading engine 4633 was a long term Canton engine, much employed as one of the station pilots at Cardiff General. (R.K. Blencowe Collection)

The Dowlais Ammonia train conveyed barrier wagons between the engine and the ammonia wagons from the early 1960s, but here the northward bound train has the payload next to the engine with two 8750 engines in charge on 13 June 1959. Note the condensation on the tank wagons. (Alan Jarvis/SLS)

A Brecon to Merthyr Goods with 9776 emerges from Torpantau Tunnel on 13 June 1959. (Alan Jarvis/SLS)

The northern portal of Torpantau Tunnel as seen from the footplate of a southbound service on 18 May 1959. (Alan Jarvis/SLS)

A view from the footplate climbing Seven Mile Bank on 18 May 1959. (Alan Jarvis/SLS)

LOCATION ANALYSIS • 93

A panorama of Torpantau, as seen from above the tunnel.

The SLS Special on 2 May 1964 runs through Torpantau still partly shrouded in fog on the Merthyr to Brecon portion of the tour, with 4555 leading and 3690 inside.
(John Hodge)

Descending the Seven Mile Bank between Torpantau and Pentir Rhiw, 7771 heads a three coach service from Newport on 28 May 1958, before trains were passed to the care of the more modern 8750 Class of Pannier. (T.B. OWEN/Colour Rail)

LOCATION ANALYSIS • 95

The Torpantau signalman is about to exchange tokens with the driver of this Brecon to Merthyr freight which continued to run after the closure of the passenger service. (Ken Mumford Presentations)

The south end of Torpantau station, which was probably the most photographed isolated place on the B&M. (Ken Mumford Presentations)

Torpantau SB and northward platform. (D.K. Jones Collection)

3706 runs into the northbound platform with a train from Newport in 1962 (D.K. Jones Collection)

2280 on the 2pm Brecon to Newport enters the double track section through Torpantau on 16 May 1959.
(D.K. Jones Collection)

7736 stands at Torpantau with the 12.10pm Brecon to Newport on 16 May 1959.
(D.K. Jones Collection)

Trains in the landscape – a pair of Pannier Tanks blasting up the last few yards of the 1 in 40 gradient to Torpantau of the loaded ammonia tanks from Dowlais. After passing through Torpantau Tunnel, the next job would be to pin down wagon brakes to safely descend the 1 in 38 of the Seven Mile Bank. The Pannier Tanks will haul the train as far as Hereford, where main line power will come on, en-route to Billingham. (D.K. Jones Collection)

9638 leaves Torpantau Tunnel with a Merthyr to Brecon goods, the first wagon conveying large coal on 7 September 1962. (Michael Roach)

LOCATION ANALYSIS • 99

Seen from the mountainside above the station 3201 leaves Torpantau with the 2.5pm Brecon to Newport on 8 September 1962. (Michael Roach)

Torpantau looking south from the north end of the platforms on 8 September 1962. (Michael Roach)

Torpantau looking north towards the station on 8 September 1962. (Michael Roach)

Ebbw's 3706 heads the 1115am Newport to Brecon calling at Torpantau on 18 April 1962. (Malcolm James)

LOCATION ANALYSIS • 101

A Newport bound train, headed by a pannier tank, climbing the seven Mile Bank.

A pannier tank working hard approaching the north portal of Summit Tunnel just east of Torpantau, through the superb scenery surrounding this section of line. The load on this section of the line was 120tons for an 8750 Class, equivalent to 4 coaches, here being 3 coaches and a horsebox. (Ken Mumford Presentations)

The 2pm Brecon to Newport, with horsebox tail traffic, is about to enter the 667yard Summit (Beacon) Tunnel after climbing at 1 in 38 from Talybont, headed by Brecon based Dean Goods 2569 on 12 August 1943. *V. Webster*/(Ken Mumford Presentations)

A late addition to the Ebbw Jct. fleet of 2251s was 2209, though it didn't survive long. Here it is seen waiting at Torpantau for down train to arrive from Brecon hauled by an 8750 Class pannier in 1962. (Ken Mumford Presentations)

LOCATION ANALYSIS • 103

A Class 37 diesel on the B & M - on an engineer's track recovery train well after closure to traffic. The diesel is seen emerging from the summit tunnel and was probably an everyday scene, when the line was being recovered in this area.

A view of the southern portal of Summit Tunnel after closure and removal of the track.
(Ken Mumford Presentations)

The interior of Summit Tunnel following closure showing how some sections were the original rock, with others of stone and brick. (Ken Mumford Presentations)

The south portal of Summit Tunnel on a lovely sunny day, situated about 200 yards from the site of the former station, now basically a muddy stream. (Ken Mumford Presentations)

Pentir Rhiw

At 36miles 79ch from Newport, the layout and SB here were brought into use in January 1897, including a short island platform 'for use on two days a week for markets at Brecon and Dowlais'. A runaway siding was provided just north of the station for the benefit of trains on the Seven Mile Bank and a loading siding for timber for several years from 1915. In 1909 the platform was made public and, in later years at least, tickets were sold by the signalman through a standard booking office hatch incorporated into the SB windows. In May 1953, the Down Distant Signal (for trains proceeding Up the bank) was made workable, so that drivers would know that it was not necessary to stop at Pentir Rhiw and the section ahead to Torpantau was clear. The single line tokens would be exchanged with the train on the move, so that it did not lose momentum, though a water stop was available a few chains on. The layout and SB were taken out of use after cessation of passenger operations in December 1962, in January 1963.

For negotiating the Seven Mile Bank, seven miles at 1 in 38, passenger trains were allowed 28-30mins (dependent on load) in the northerly direction and 15mins running south. Freight trains were allowed 44mins running north and 35mins running south, though the latter would be increased for stopping to pin down brakes as required by weather and load. Communication between the driver and signalman was critical. Three crows on the whistle meant the train was under control and could be allowed to proceed down the bank, so the left hand signal at Pentir Rhiw could be lowered. A long blast on the whistle indicated Danger and the whole train was diverted into the runaway siding, for which the right hand signal was lowered.

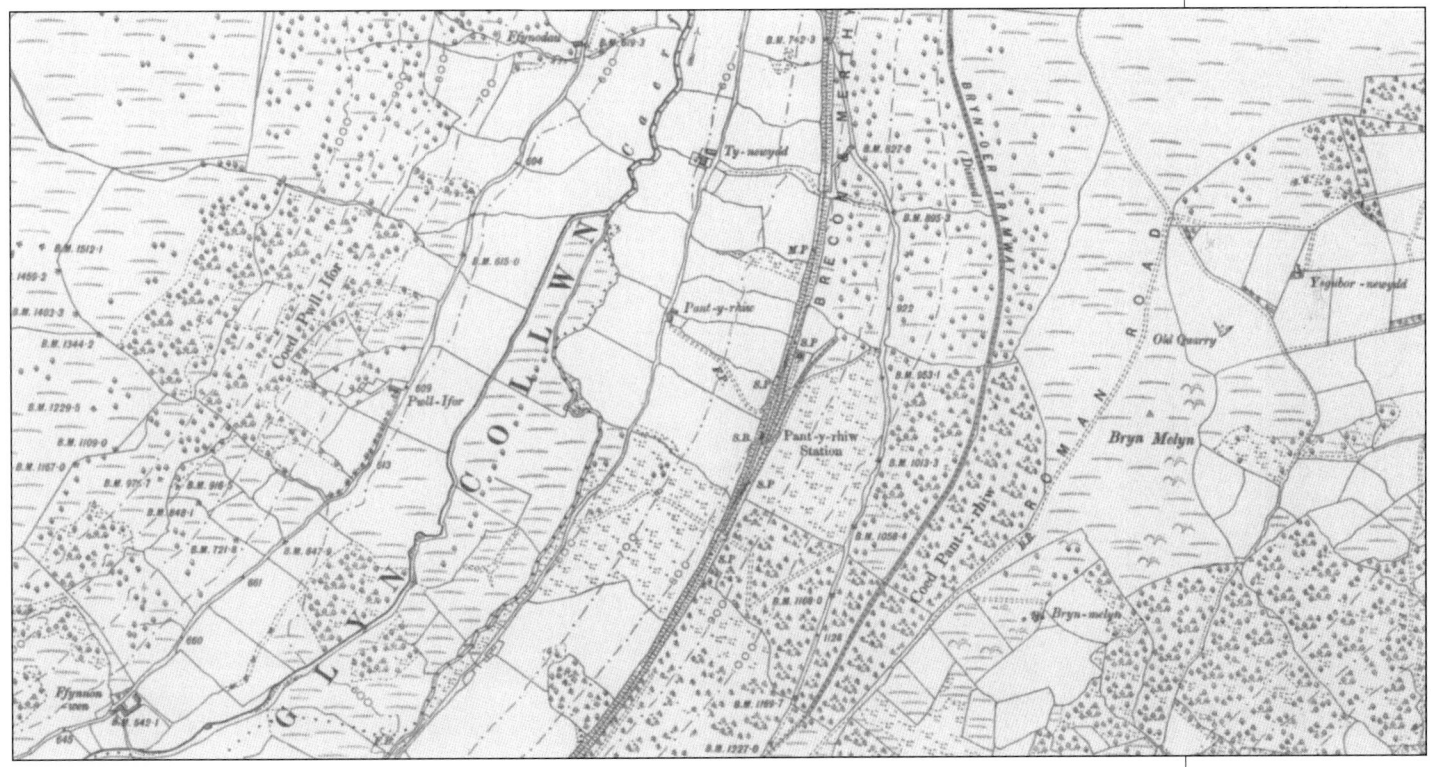

OS Map of the area around Pentir Rhiw (Pant-y-Rhiw) at the turn of the twentieth century. (National Library of Scotland)

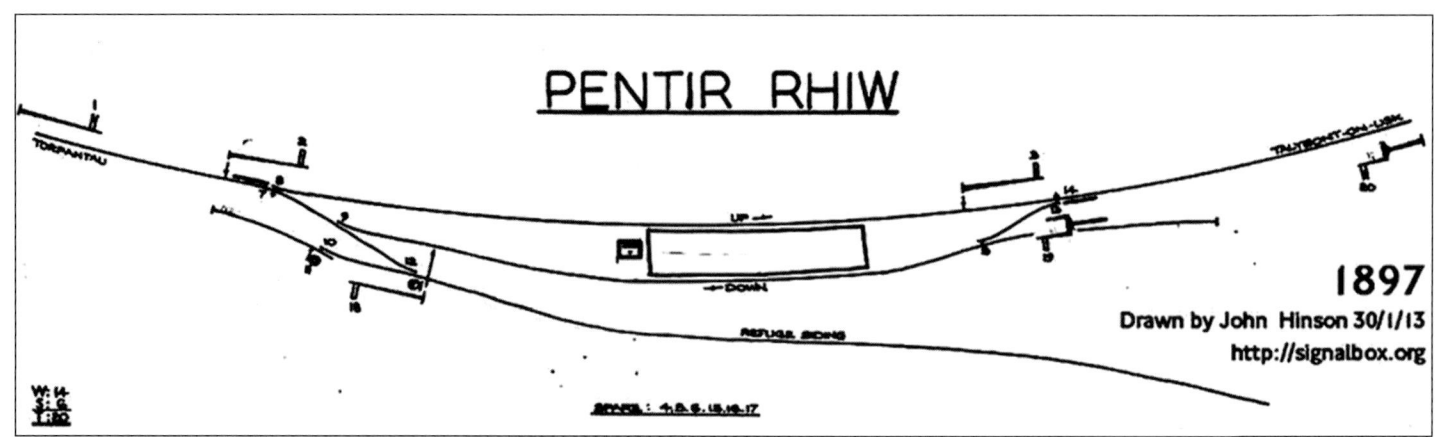

A diagram of the track plan at Pentir Rhiw. (John Hinson)

The small island platform of Pentir Rhiw with the northbound line in the foreground. (GWT)

The southbound track at Pentir Rhiw station on 7 September 1962. (Michael Roach)

As a crossing point, tokens needed to be exchanged between driver and signalman, with the driver of this southbound train presenting the token from Talybont to Pentir Rhiw in exchange for that from Pentir Rhiw to Torpantau. (Ken Mumford Presentations)

46524 approaches Pentir Rhiw with the 2.10pm Brecon to Newport on 10 July 1958. (R.M. Casserley)

In the same spot the return Dowlais Ammonia tanks hauled by Hereford's 8722 in June 1960. (D.K. Jones Collection)

LOCATION ANALYSIS • 109

Approaching Pentir Rhiw from Torpantau, 2240 is working back home northwards with the Bassaleg to Brecon Goods on 16 May 1962. (A. Cooke/Colour Rail)

Seen from a northbound service, 9425 is standing at the platform, waiting access to the single line south of Pentir Rhiw on 18 July 1959. (R.E. Toop)

3747 gets up steam to tackle the Seven Mile Bank as it makes the stop at Pentir Rhiw with the 12,10pm Brecon to Newport service on 7 September 1962.
(Michael Roach)

The view past the SB and along the platform to the north on 7 September 1962.
(Michael Roach)

LOCATION ANALYSIS • 111

The view from a southbound train on leaving Pentir Rhiw on 27 September 1958. (R.J. Buckley/Initial Photographics)

A view from the signal box at Pentir Rhiw of an engine and brakevan heading south. (H.B. Priestley)

9776 running south with a brakevan south of Pentir Rhiw alongside Talybont Reservoir, constructed by Newport Corporation in the 1920s. (H.B. Priestley)

The layout south of Pentir Rhiw, showing the runaway siding on the right. The procedure was for the road towards Brecon to be set for the runaway siding until the approaching train had come to a stand at the home signal, and whistled to indicate that it was under control. The road would then be set for the train to come into the loop, and the signal pulled off. A siding going off the runaway siding had been used for loading timber in times past. (H.B. Priestley)

LOCATION ANALYSIS • 113

Approaching Pentir Rhiw from the south on 13 June 1959.

South of Pentir Rhiw showing the main lines and runaway siding in 1958.
(D.K. Jones Collection)

Having made the station stop at Pentir Rhiw, 3747 coasts towards the water stop with the 12.10pm Brecon to Newport on 7 September 1962. (Michael Roach)

3747 taking water working the 12.10pm Brecon to Newport south of the SB and alongside the start of the runaway on 7 September 1962. (Michael Roach)

With tanks replenished, 3747 heads off for Torpantau. (Michael Roach)

3706 heads north with the 8.3am Newport to Brecon making the Pentir Rhiw call on 7 September 1962. (Michael Roach)

In the hills around Pentir Rhiw 3201 heads for Torpantau on 7 September 1962. (Michael Roach)

In the ninth month after the withdrawal of the passenger service, 9676 heads south approaching Pentir Rhiw station with a Brecon to Merthyr freight, composed of vans and empty coal wagons. By then the signal box and passing loop had been taken out of use. (Ken Mumford Presentations)

A prelude to the big freeze at the end of December, and January 1963 with the end of the B&M passenger service only days away, a snowy scene at Pentir Rhiw as a southbound train faces another few miles of the 1 in 38 climb. (Ken Mumford Presentations)

LOCATION ANALYSIS • 117

A train descending the Seven Mile Bank has come to grief at the entrance to the Pentir Rhiw loop. (Ken Mumford Presentations)

The view from near the top of the Seven Mile Bank with Talybont reservoir below. (Ken Mumford Presentations)

Talybont-On-USK

The station was a fairly standard two-platform structure at the bottom of the Seven Mile Bank, 40miles 20ch from Newport. A small goods yard of three interlinked sidings was provided on the Down side of the line. In early days, this included an engine shed, provided for banking engines to assist trains up the 1 in 38 gradient, but it was later found more convenient to locate these at Talyllyn where a considerable amount of shunting took place, so the engine shed became used as a goods shed. By December 1896, a Down Refuge Siding had been added at the south end of the layout, with a further short siding off it, incorporating cattle pens.

An over-run siding, some half mile long, was provided at the north (Brecon) end of the layout and by 1927 a loop siding had been added at the northern end of this, believed to have been used for the unloading of roadstone for the local County Council. Part of the goods yard and a cross over were taken out of use in January 1963; the rest of the layout, apart from a single line, together with the SB, followed in the September.

All tank engines took water at Talybont, but for tender engines, taking water at Bargoed sufficed.

After closure of the railway, the station buildings were adapted and added to by a consortium of local authorities for use as an outdoor pursuits centre, which has since ceased due to financial stringency.

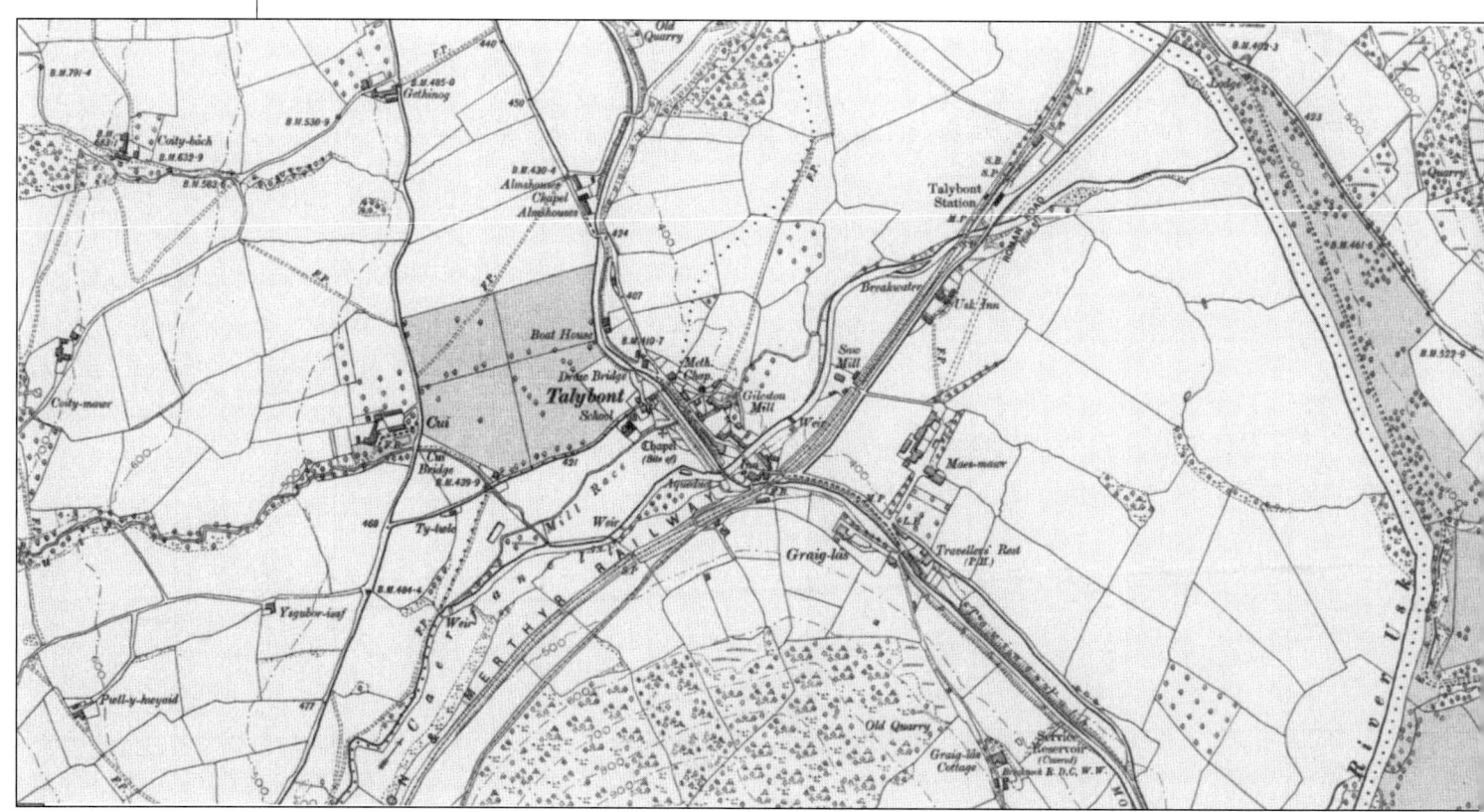

Above: OS Map of the area around the B&M line at Talybont on Usk at the turn of the twentieth century. (National Library of Scotland)

Opposite above: An early 19 C. depiction of Talybont on Usk where the Bryn Oer Tramroad met the Brecknock & Abergavenny Canal. (Ken Mumford Presentations)

Opposite below: Talybont on Usk in 1904. (LGRP)

LOCATION ANALYSIS • 119

Messrs. Dixon & Overton ...to deliver in Boats at the Public Wharf at Brecon, Sixty or more tons of coal per diem at Twelve Shillings per Ton and any quantity of Lime, not less than a Boat Load per diem likewise in Boats at the same place for Fourteen Pence per Barrel
[25 April 1815]

Flux was used in a blast furnace to cause the impurities in the ore to melt readily and become fluid at furnace temperature. Limestone was an ideal flux and in 1843-74 it took 14-15 cwt of limestone to produce 1 ton of iron.

George Overton's limekilns at Talybont on Usk of 1815. Alternate layers of limestone and coal were charged from the top and lime drawn from the arches at the bottom. It was then transported in barrels by narrow boats.

A double headed down train in B&M days, possibly in 1904. This is probably the early afternoon train from Brecon which divided at Pontsticill. The Merthyr portion would proceed in charge of the leading saddle-tank engine, while the following 2-4-0 tank would head the remaining coaches to Newport.

Talybont on Usk looking north on 20 August 1961.
(Lens of Sutton)

LOCATION ANALYSIS • 121

The scene so often seen during winter with the up platform covered in snow in December 1962. (John Hodge)

Talybont on Usk looking south on 18 April 1962. The start of the 1 in 38 climb to Torpantau is clearly visible ahead. (H.B. Priestley)

The station exterior on 20 August 1961. After closure, the station buildings were adapted and added to for use as an outdoor pursuits centre by a consortium of local authorities, but in austerity, this use has now ceased. (Lens of Sutton)

Merthyr pannier 4690 accelerates through Talybont with a returning freight from Brecon to Merthyr. (Colour Rail)

LOCATION ANALYSIS • 123

A returning Brecon to Merthyr freight, hauled by Merthyr panniers 9676 and 4690, has paused at Talybont on Usk (385ft. above sea level) for both engines to quench their thirst, prior to facing the onslaught of the Seven Mile Bank at 1 in 38 to the Summit or Beacon Tunnel to Torpantau (1314ft. above sea level).
(Ken Mumford Presentations)

Brecon's 2280 with the 2.10pm Brecon to Newport on 18 May 1959. The engine was withdrawn a few months later.
(R.O. Tuck)

Seen alongside the down platform Ebbw's 2247 has the same train as in the previous picture on 13 October 1962. (W.G. Sumner)

Four engineer's clerestory roofed coaches were stored south of Talybont station in 1959, seen here as a down train regains the single line on 16 July 1959. (H.C. Casserley)

LOCATION ANALYSIS • 125

Running into Talybont from Pentir Rhiw on 19th June 1962 with wagons of timber in the up sidings. (Garth Tilt)

3634 runs into Talybont on Usk over the river bridge at the south of the station with a Newport to Brecon service on 8 September 1962.
(Michael Roach)

Seen from the north end of the down platform, 3700 has the 8.3am Newport to Brecon at the up platform on 12 October 1962, showing also the layout at that end of the platforms.
(W.G. Sumner)

3700 at the Up Platform with the 8.3am Newport to Brecon service on 12 October 1962.
(W.G. Sumner)

Talybont on Usk SB at the north end of the up platform. Talybont SB dated from 1892, being a typical McKenzie & Holland brick-built structure, housing a 17 lever frame. The stone-built water column alongside is worthy of note.
(Ray Caston Collection)

A local cow makes her daily inspection of the platform on 7 July 1962.
(Alan Jarvis/SLS)

The substantial stone building seen here was originally constructed as an engine shed to house banking engines for the climb to Torpantau. It was opened around 1863 but by 1898 was in use as a goods shed, it having been found more convenient to have an engine shed at Talyllyn Jct., where a substantial amount of freight was exchanged. It was later used as a non rail-connected storage unit by a local haulier, as seen here in July 1962. (Alan Jarvis/SLS)

2247 starts away from Talybont on Usk with the 2.15pm Brecon to Newport on 7 July 1962. (Alan Jarvis/SLS)

The final train over the line, the SLS Special on 2 May 1964, seen running in on the Brecon to Dowlais leg of the tour with 4555 leading and 3690 inside. By then the signal box and loop had been taken out of use. (John Hodge)

The train at the south end of the station with assembled photographers. (John Hodge)

An elevated view of the special and accompanying throng during the stop at Talybont. (Jeff Stone)

LOCATION ANALYSIS • 131

The SLS Special preparing to depart Talybont on Usk on its way from Brecon to Dowlais on 2 May 1964.

2½ years after closure and little has changed in this view from south of the river bridge on 20 June 1965. The SB had closed, signal arms removed and points clipped for through running between Brecon and Pontsticill Jct. in September 1963.
(Jeff Stone)

The signals are disconnected but this shot on the same date after closure affords a view of the cattle pens on the down side. (Jeff Stone)

Talyllyn East Junction

This was the point at which services to Mid-Wales deviated from the main B&M line to avoid the Junction station. In the normal course of events, there were few such services but pre-1939, this was where the Cardiff/Treherbert to Aberystwyth and Barry to Llandrindod Wells took the chord line from Talyllyn East to North Junctions en route to the Mid Wales line.

Talyllyn East Junction was where trains could switch from the Brecon line to the Mid Wales and Hereford lines and vice versa. It was used in the past by the services between Merthyr and Aberystwyth and later by the Dowlais ICI Ammonia Tank traffic en route to Hereford. This view after closure on 4 August 1963 shoes the situation after the signal box and junction had closed, with only the through line to Brecon on the left in use. (Garth Tilt)

LOCATION ANALYSIS • 133

The Talyllyn East Jct. signalman takes the token from the driver of a Newport to Brecon service on 13 October 1962.
(Garth Tilt)

This view shows the junction signal coming off the line from North Junction and shows 46522 running light, possibly to turn on 8 June 1962.
(Alan Jarvis)

134 • RAILWAYS & INDUSTRY ON THE BRECON & MERTHYR: MERTHYR-PONTSTICILL JUNCTION-BRECON

The building of the old Mid-Wales railway station at Talyllyn North Jct. which closed in 1878. It is said to have been subsequently used as a private dwelling, and is seen here in 1958.
(D.K. Jones Collection)

Between Talyllyn East and West Junctions, the 11.15am from Newport approaches Talyllyn station running past the small goods yard on the left. The first station, called Brynderwen, opened somewhere between here and the later Talyllyn Jct. station, on 23 April 1863
(D.K. Jones Collection)

Talyllyn Junction

The original B&M station here was called Brynderwen and was sited slightly to the south of the replacement Talyllyn Jct. station, the latter at 43miles 18ch from Newport. A small goods siding was located immediately south of the station, just one siding with loading bank and cattle pens.

As the main interchange point between the B&M, Mid Wales (later Cambrian) and Midland Railways, much shunting of goods vehicles took place at Talyllyn, facilitated by the loop lines and sidings. Railway Clearing House number-takers were stationed here to keep track of the various transfers.

In the centre of the triangle of lines, the B&M opened an engine shed in 1863, partly to replace that at Talybont. It seems to have had a chequered career, being later used as a wagon repair depot, then back to use as an engine shed. By 1903, the major part was a carriage shed, with a lean-to single road engine shed alongside. All was taken out of use in late 1922, following the Grouping, though it was left in situ and a 1940 photograph shows it gently collapsing into ruins.

With the complete closure of the line towards Hereford and Moat Lane at the end of 1962, all lines were taken out of use on 13 January 1963, except the single line to Brecon, together with the East and West SBs.

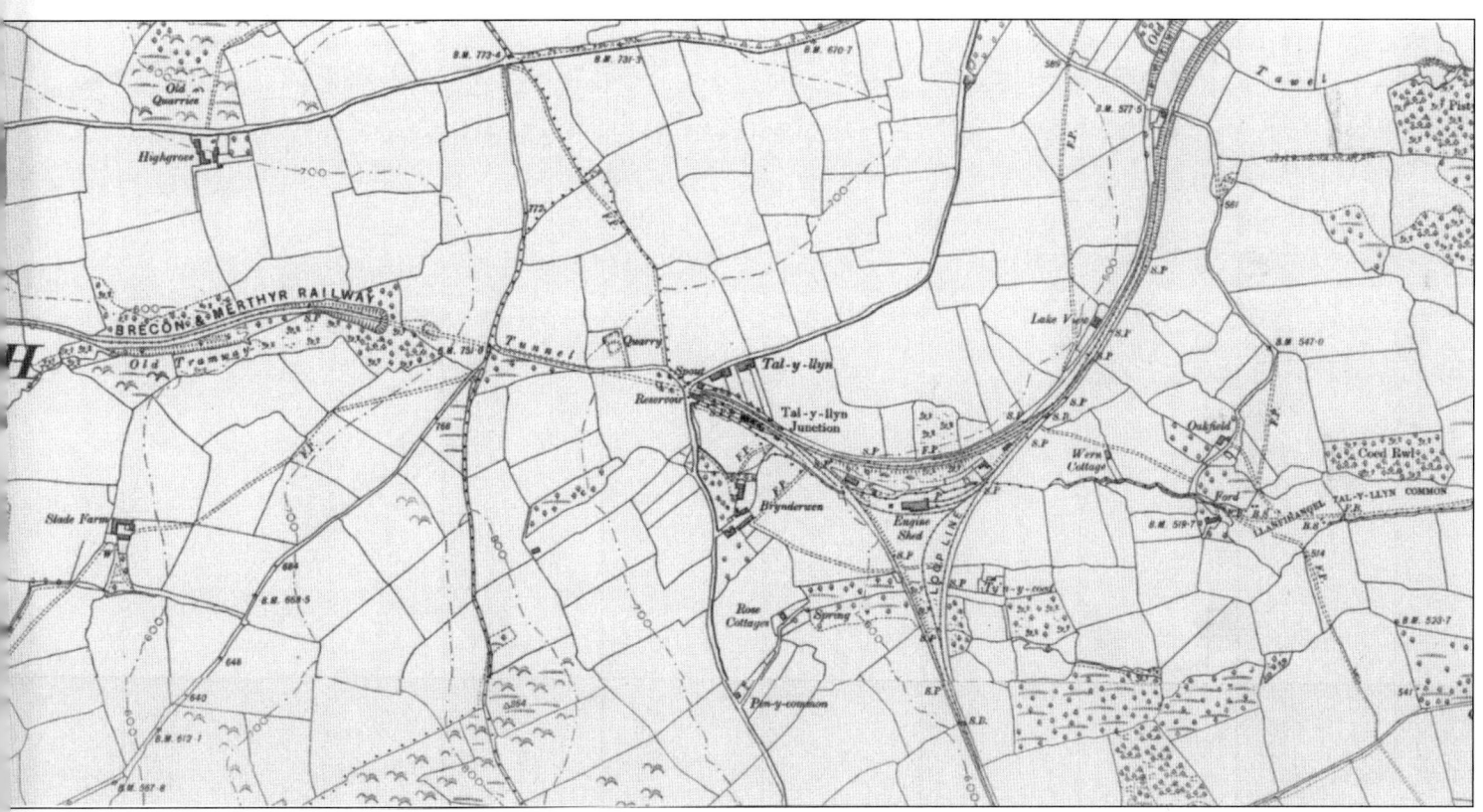

OS Map of the area around Talyllyn (Tal-y-llyn) Junction at the turn of the twentieth century, showing the East Junction giving access to the Mid Wales line south of the station and the triangular junction just east of the station with the tunnel on the line on to Brecon. (National Library of Scotland)

Views of Talyllyn Station, the first looking north to the tunnel at the end of the platforms on 18 May 1959. (Alan Jarvis/SLS)

A view of the station on 14 April 1962. *R.Patterson/*(Colour Rail)

LOCATION ANALYSIS • 137

Crossing the line was achieved using the barrow crossing at the south end of the station, under the eye of the signalman. In view of the large number of trains involved, this had to be a well supervised procedure, as here on 18 June 1951. (T.C. Cole)

The SB, signal and junction at the south end of the platforms, showing the beginning of the Cambrian line up platform beyond. This allowed two trains from Brecon to be present at the same time, allowing connections to Hereford and Mid-Wales from one Newport to Brecon train As will be seen in the following photographs, there was often a lot of passenger interchange here, and it was not unknown for some trains to carry on to Brecon practically empty. (Michael Roach)

138 • RAILWAYS & INDUSTRY ON THE BRECON & MERTHYR: MERTHYR-PONTSTICILL JUNCTION-BRECON

The down platform in the snow in December 1962. (John Hodge)

Looking north up the Up Platform showing the station nameboard. (GWT)

LOCATION ANALYSIS • 139

Looking north along the two platforms from the south end of the down on 8 September 1962. Amongst the facilities offered here was a licensed refreshment room.
(Michael Roach)

Looking south from the north end of the Down Platform on 8 September 1962.
(Michael Roach)

The north end of the station looking to the tunnel. (SLS)

The tunnel notice at the north end of the up platform as on 29 September 1962. The plaque commemorates the fact that Talyllyn Tunnel is one of the oldest railway tunnels in the world, and was for many years the oldest still in use. (W.G. Sumner)

LOCATION ANALYSIS • 141

The Indicator Arms at the station seen on the final day of service. (Malcolm James)

Ebbw's 2227 waits at the down platform with the 12.10pm Brecon to Newport train as a Cambrian 0-6-0 runs into the up platform with a train from Mid Wales on 11 September 1951. The stone-built Talyllyn West Jct. SB is prominent; note the passengers waiting to cross the line (H.C. Casserley)

142 • RAILWAYS & INDUSTRY ON THE BRECON & MERTHYR: MERTHYR-PONTSTICILL JUNCTION-BRECON

The final months of working saw the service in the hands of 2251s and Panniers. Here two views of Ebbw's 2247 at the down platform with the 2.5pm Brecon to Newport on 13 October 1962. (W.G. Sumner)

Until the early 1950s, Dean Goods 2301s were used on the B&M services, as here with Brecon's 2401 at the down platform on 3 May 1951. (R.C. Riley/Transport Treasury)

Below and overleaf: Three views of 2298 on the 2.5pm Brecon to Newport, the first crossing a 465XX on the 12.30pm Builth Road to Brecon at the Up Platform, on 29 September 1962. (W.G. Sumner)

LOCATION ANALYSIS • 145

A group of the newer Panniers worked the Brecon services from Newport, such as 3638 seen at the Up Platform with the 11.15am Newport to Brecon on 3 May 1959. (S. Rickard/DKJ J&J Collection)

Below and overleaf: Three views of 3747 taken from the north end of the down platform as a train leaves on the final stage of its journey into Brecon, the last approaching the Tunnel with the end in sight. (Michael Roach)

146 • RAILWAYS & INDUSTRY ON THE BRECON & MERTHYR: MERTHYR-PONTSTICILL JUNCTION-BRECON

LOCATION ANALYSIS • **147**

Two views from the south end of the station on 29 September 1962, the first showing the junction as 3747 approaches with the 8.3am from Newport, as 46523 waits at the Cambrian line platform with the 10.25am Brecon to Hereford. In the foreground, the sole goods siding, complete with cattle pens, can be seen on the right. (Both W.G.Sumner)

Brecon's 2236 with the 2.10pm Brecon to Newport threads the shadows of the down platform on 5 November 1960. (M.B. Warburton)

Cambrian 0-6-0s were an everyday sight until the early 1950s as here with 896 on what may be a 3 coach Newport service at the down platform on 19 November 1948. (T.C. Cole)

887 prepares to leave with a train to Moat Lane on 26 March 1951. (Ian L. Wright)

Another common sight on the Hereford services in the early 1950s were the former Lancashire & Yorkshire 3F 0-6-0s, such as Hereford's 52525 seen at Talyllyn on 3 August 1951. (R.C. Riley/Transport Treasury)

Most of the Mid Wales and Hereford services and the two Brecon-worked Newport services in the later 1950s were covered by the Standard 2MT 2-6-0 465XXs as here with 46523 emerging from the tunnel with the 10.25am Brecon to Builth Wells on 29 September 1962. (W.G. Sumner)

The signalman exchanges tokens with the driver of 11.15am Newport to Brecon who surrenders the token from Talybont and receives one through to Brecon, as 46515 awaits departure at the down platform with the 1.40pm service to Mid Wales. (R.E. Toop)

The last train from Brecon to Newport left at 6.15pm on 29 December 1962 and was recorded in the dark at Talyllyn by my friend Malcom James, double-headed by 9776 and 3700, no doubt conveying extra stock to help clear such from Brecon.

A double-headed troop special from Sennybridge bound for Hereford with 2240 leading running into Talyllyn on 9 June 1962.
(R. Patterson/Colour Rail)

Talyllyn West SB in the snow.
(Mike Morton Lloyd)

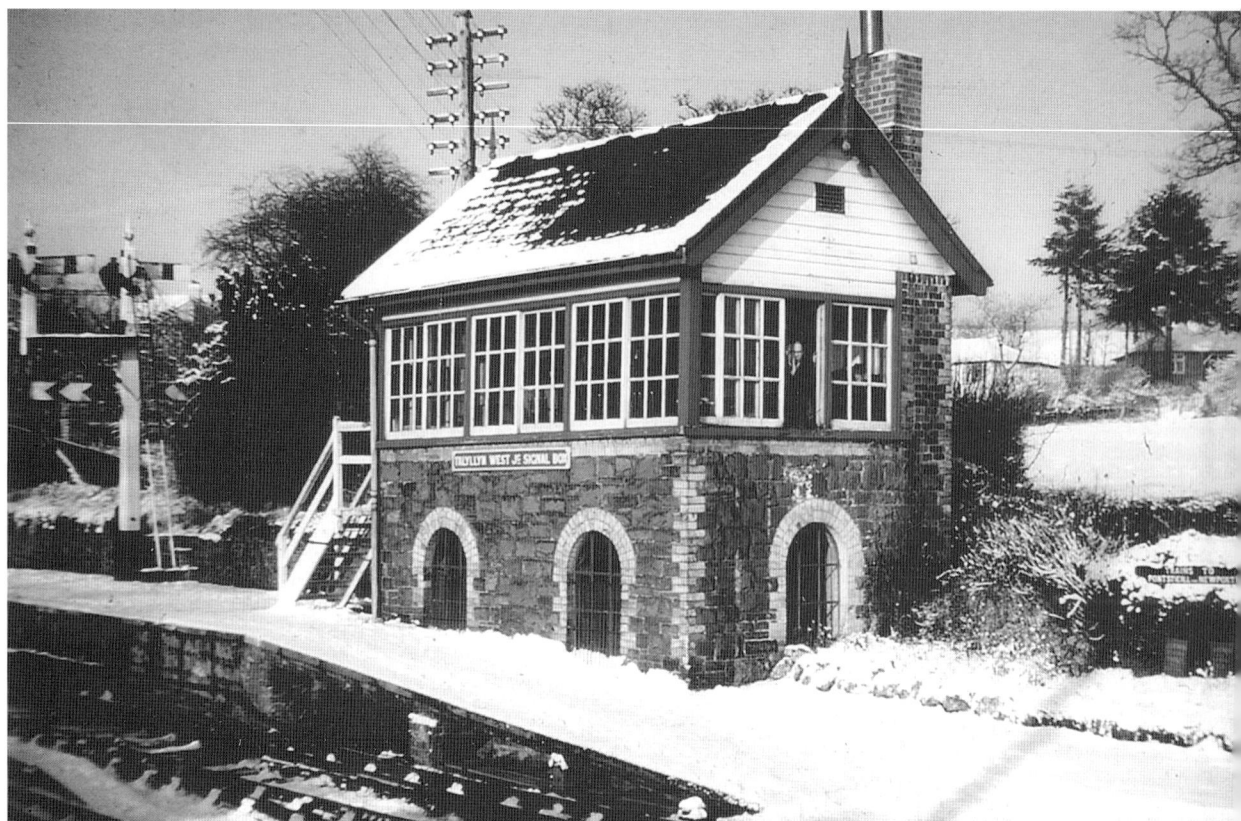

Talyllyn station is now a private residence and has been landscaped into this scenic garden.
(Ken Mumford Presentations)

Groesffordd Halt

At 45miles 20ch from Newport, Groesffordd Halt was opened by the GWR on 8 September 1934 and remained in use until the closure of the service on 31 December 1962. Located on the single line, it was a short platform on the south side of the line.

A view looking north along the single platform. (Lens of Sutton)

A 465XX calls at the halt running tender first on 8 September 1962. (Michael Roach)

Opened by the GWR in 1934, two views of the halt from both directions, the first in 1951, the second on 8 September 1962. (Stations UK, Michael Roach)

Brecon Yard & Shed

Brecon Jct.SB, which controlled the entrance to Brecon Yard, was approx.46miles 40ch from Newport. The box was closed in November 1931 and replaced by Brecon Yard GF. In Brecon Yard, the first passenger station at Watton was opened in April 1863 and closed in March 1871 with the transfer of passenger facilities to Free Street, the Watton site then used as offices, though part of the former platform was used as cattle pens. The track running past the former platform performed the very useful function of joining the yard with the passenger station, the usual route for engines to and from shed.

To the left of this track was situated the five road goods yard, complete with shed. Left again was the loco yard with firstly the single road shed of the Cambrian Railways. This was demolished c.1934, leaving the stores building with water tank above and loco pits for further use. The B&M shed alongside had two roads under cover, with a further siding on each side of it and was refurbished in 1934. It would appear to have been re-roofed at this time, judging by earlier and later photographs. There were two further sidings alongside and to the rear of the shed. There was also a lower yard with four or five more sidings, home to a sawmill and timber yard in 1904, though how much of the yard these occupied is unclear. To the west was a cart weighbridge and offices, convenient to the road access to the yards. All in all, the yards here were the largest complex on the B&M.

Brecon Goods Yard with 36XX shunting in the late 1950s. (Colour Rail)

The original station at Brecon was at Brecon Watton before operations were moved to the much larger Free Street, opened in March 1871. The platform and buildings remained at Watton in use as offices and stores, while the line running past the old platform was a useful link with Free Street for engine purposes. The higher level carriage sidings can be seen in the background, the goods yard and shed to the left, whilst to the right of the former station buildings are cattle pens. These two views are first on 9 May 1953 and second in September 1962. (T.J. Edgington, Michael Roach)

LOCATION ANALYSIS • 157

These two views give an idea of Brecon shed in its original condition before it was "reconditioned" by the GWR, apparently re-roofed, at which time it was probably that the ex-Cambrian shed was demolished, though the water tank, stores and pit were retained. In B&M days engines were turned out in immaculate condition as here with No. 12, probably awaiting a run to Newport c.1922. She did not last long after the grouping and was withdrawn, still based at Brecon, in February 1923. (LCGB/NRM)

In not quite so good condition, B&M 38 at the shed on 12 August 1913. Following the Grouping, she became GWR 332 and was allocated 423 though was never renumbered. She spent time at Severn Tunnel Junction shed, shunting the yards, before being withdrawn in December 1949. (LCGB/NRM)

B&M 19 was also a Brecon engine, but was withdrawn in January 1923. Note the Cambrian Collieries wagon on the track alongside. (LCGB/NRM)

B&M 28, an 0-6-0ST then based at Brecon, but moved to Bassaleg after the Grouping as GW 2172 and withdrawn in 1928. (LCGB/NRM)

These former Midland Railway 0-6-0Ts were based at Brecon by the LMS mainly for working Midland traffic between Swansea and Worcester etc. Their presence was significantly reduced when this long distance traffic ceased after the Grouping, by about 1930. The roof of the ex-Cambrian Railway shed is visible above the boiler of engine 1959, the photos showing MR 1629 and 1959 at the shed c.1922. (Both LCGB/NRM)

Midland Railway 3F 0-6-0 3462 with crew at the shed c.1922. (LCGB/NRM)

Former L&Y 0-6-0 12143 at the shed in April 1947, when they worked the Hereford service. (R.K. Blencowe)

Brecon was a stronghold for 0-6-0s and the GWR based several of the Dean Goods 2301 Class there with six based there at Nationalisation on 1.1.48. 2386 is seen at the depot on 6 September 1936 (W.A. Camwell/SLS)

LMS and GWR 0-6-0s standing side by side at the depot as here with LMS 3600 and GWR 2342 alongside another on 6 September 1936. (W.A. Camwell/SLS)

517 Class 0-4-2T 848, based at Builth Wells, at Brecon Shed on 6 September 1936. (W.A. Camwell/SLS)

2177 at Brecon Shed on 17 April 1927.

Brecon based 517 Class 0-4-2T 522 at Brecon Shed on 1 May 1929. (L.A. Cockell)

A shed view in 1950 featuring former LMS 52414 and GW Pannier and 2301 Classes. (SLS)

Brecon shed in April 1954 with panniers 3767 and 3638 and 2MTs 46518/22. (Photomatic)

A shed view in September 1962 with 3201 and two 465XXs. (Michael Roach)

3714 stands outside the shed in September 1962.
(Michael Roach)

2243 of Ebbw Jct. at Brecon shed.
(A. Scarsbrook/Initial Photographics)

Ebbw Jct.'s 2247 and 46516 at Brecon shed. (A. Scarsbrook/Initial Photographics)

A view across the shed taken from the high-level lines on 12 June 1962 with 2218/40 and 46518/07/22 visible. (H.B. Priestley)

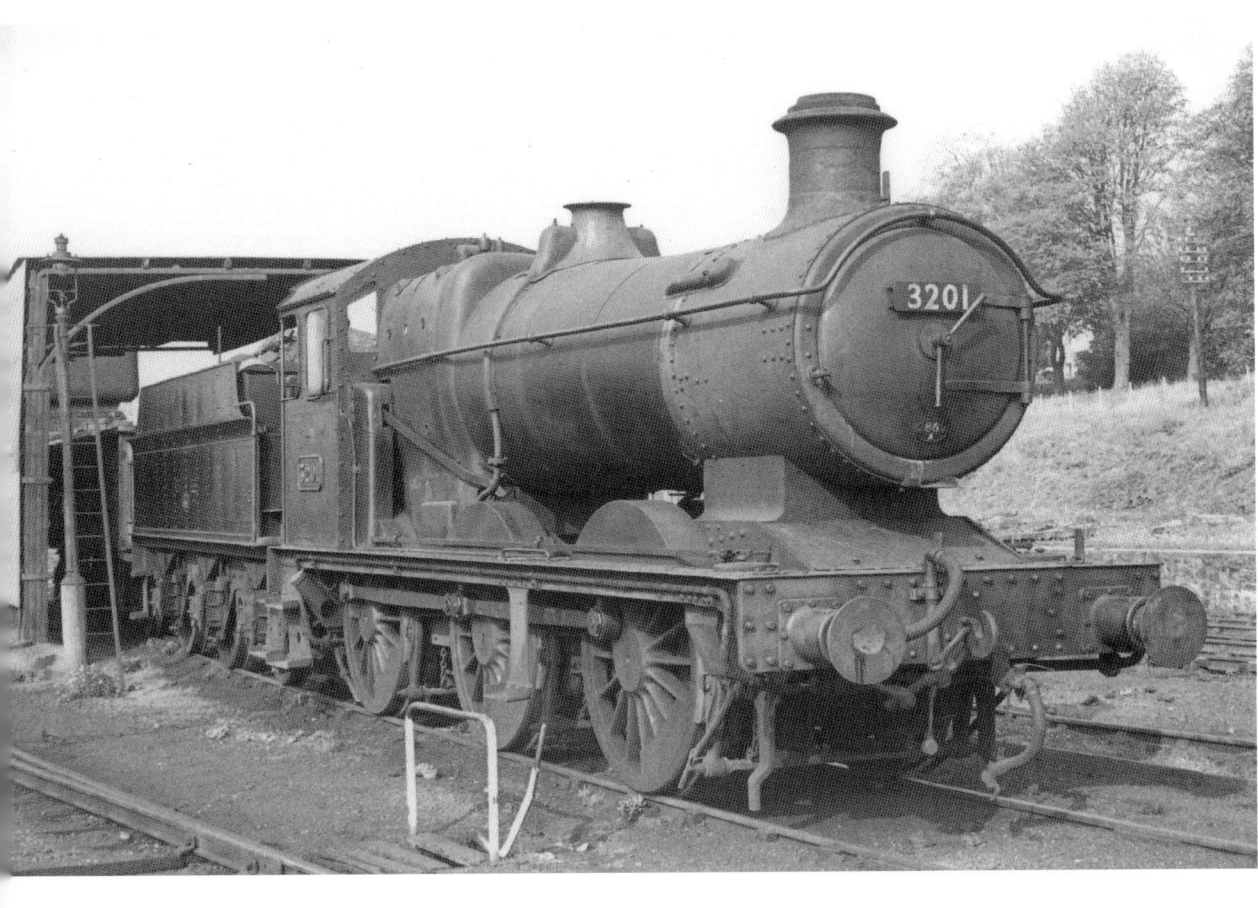

Two views of Ebbw Jct's 3201 at Brecon Shed in 1962.
(A. Scarsbrook/Initial Photographics)

A panoramic view of Brecon yards, in the twilight period between withdrawal of passenger services in December 1962 and complete closure in June 1964. On the left is the incline to the lower yard, obviously long out of use. Then we see the upper yard, with the closed engine shed in the centre, and to the right, the goods yard & shed still in use. (Jeff Stone)

Brecon Station (Free Street)

At 47miles 6ch from Newport, the new station was opened in March 1871 to replace Watton which was quite inadequate for the developing service. By 1904, there was double track from Mount Street on the Neath & Brecon through the station to Brecon Jct. The main platform supported the station buildings and there was a loop platform and two carriage sidings on the Down side. Off the south end of the main platform was a bay platform and a siding to a turntable, small coal stage and pit. It was notable that the only shelter on the platforms was a small awning directly outside the substantial main building. The layout can be seen in the various photographs following.

It had been intended to house all the company's officers here, but over the years, it was found more convenient for them to be based elsewhere. The Traffic Manager and Accountant (to deal with the growing coal traffic) and the Engineer had offices at Newport (see photograph in the first book in this series), the Locomotive Superintendent at Machen and the Company Secretary in the City of London, leaving only the Storekeeper to be housed at Brecon.

A major change to the layout was made in 1931, when the double track from Mount Street to Brecon Station and on to Brecon Jct. was taken out of use, though much was retained as sidings. Brecon Jct. SB was closed, leaving Brecon Station SB controlling the area with a new 45-lever frame. Brecon Yard and Mount Street were accessed by Ground Frames, released by the appropriate single line tokens. This layout lasted until the withdrawal of passenger services in December 1962, after which the layout was left in situ, being used by the last train, an SLS Special train, on 2 May 1964.

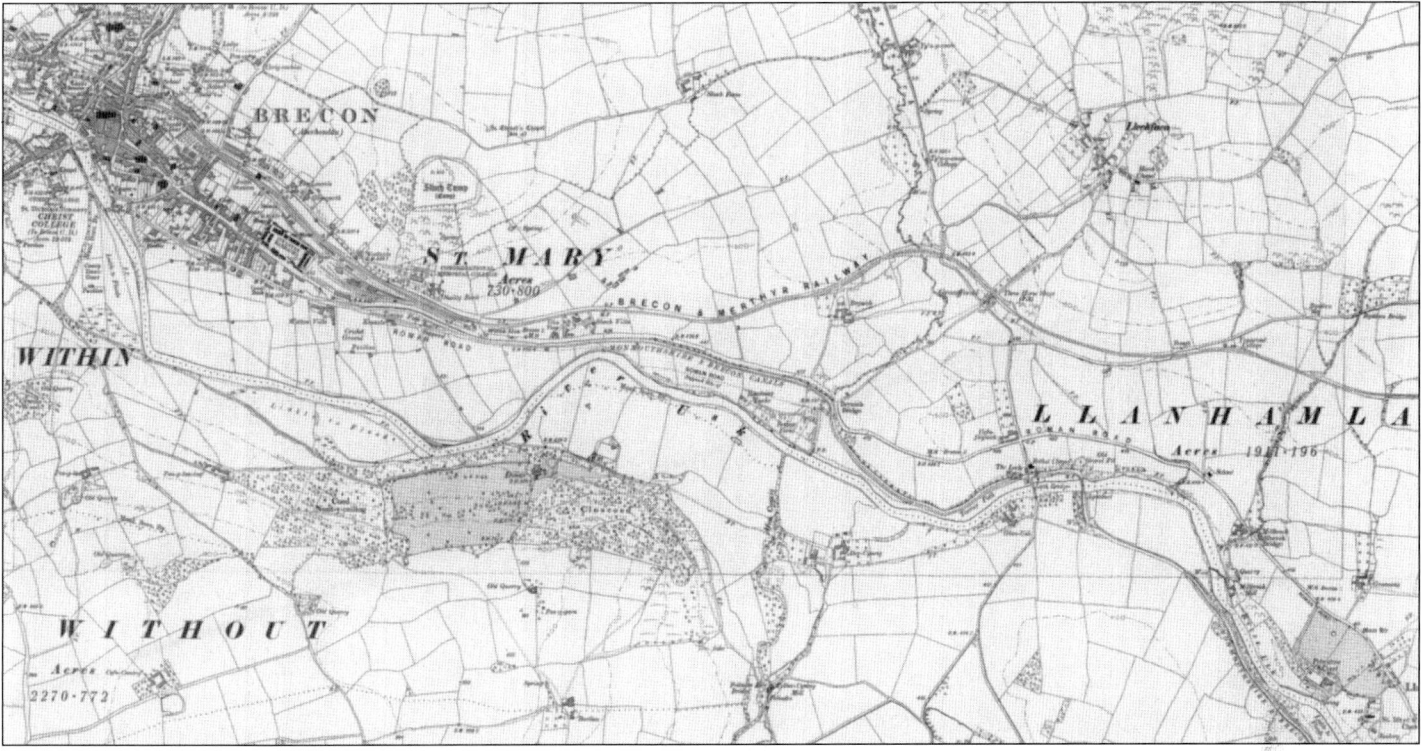

OS Map of the area between Brecon and Talyllyn showing the B&M Railway, Monmouthshire & Brecon Canal and the River Usk, all running relatively parallel east of the town. (National Library of Scotland)

OS Map of the railway approach and the town of Brecon as at the turn of the twentieth century., showing the original station at Watton and other installations in the area east of the station. (National Library of Scotland)

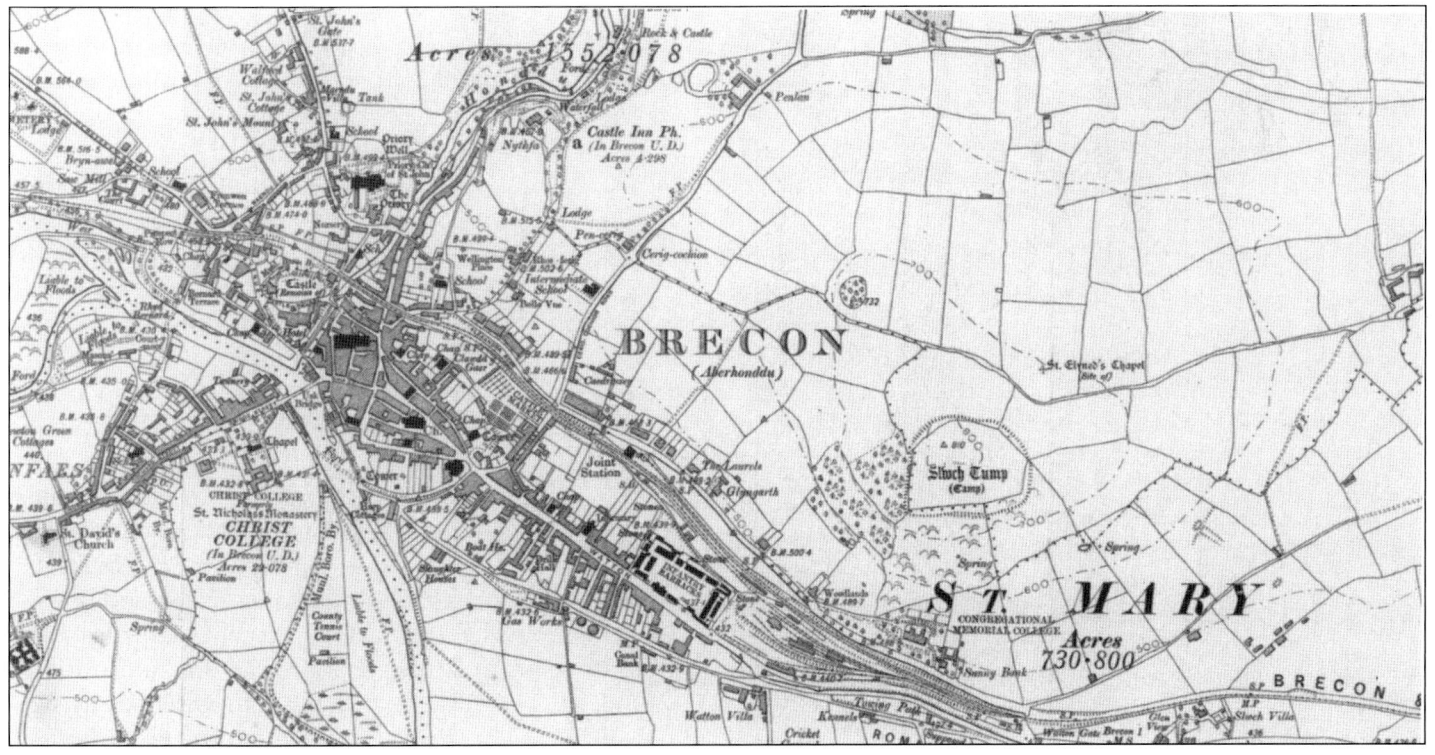

Brecon station in the early twentieth century. (E.R. Mountford Collection/Great Western Trust)

The eastward departure platform in the early twentieth century. with a double-headed train waiting to depart led by B&M 23. This is probably the early afternoon departure to Merthyr and Newport which divided at Pontsticill, the saddle tank taking the front Merthyr portion and the 2-4-0 working the remainder to Newport. (Ray Caston Collection)

Possibly another view of the afternoon departure to Merthyr and Newport which split at Pontsticill affording a better view of the rolling stock, with a full brake behind the engines. (R.K. Blencowe Collection)

A typically busy scene in the early years of the twentieth century, possibly in the early evening when the last longer-distance trains of the day all left within a few minutes of each other.

B&M 43 is turned at the station turntable, a practice always observed on the Newport services because of the Seven Mile Bank and the need to have the engine working chimney first. 43 became 1113 under the GWR and then 428, ending her days at Port Talbot shed and being withdrawn in August 1950.

Brecon free Street on 8 July 1952. A marked contrast to the previous busy scene, though there are people on the platform. The lack of any protection from the weather, apart from the small awning near the station buildings, is very evident (T.C. Cole)

In the later years, there was some through coach working between the Neath and Newport routes. Here on 15 September 1962, the rear of the 4.20pm from Neath can be seen at the main platform and this will form the 6.15pm to Newport. (Robert Darlaston)

With all platforms occupied in another Brecon "rush hour", 52525 stands in the Bay with a service to Hereford on 11 September 1951. Note the dissimilarity in the bracket signals, the modern steel bracket on the left contrasting with the older version on the right. (H.C. Casserley)

Brecon station layout looking south from the ends of the platforms in the early snow of December 1962. (John Hodge)

A Hereford to Brecon runs into the station behind LMS 0-6-0 12428 on 1 August 1947. (T.C. Cole)

Another Hereford service on 2 February 1948 with LMS 12105. The ex-L&Y 0-6-0s were used for a number of years from the mid-1930s onwards, until ousted by the arrival of the standard Class 2 456XX 2-6-0s.
(F.K. Davies)

Cambrian 0-6-0 844 with either a Mid Wales or Newport train in the bay on 28 January 1953.
(F.K. Davies)

Dean Goods 2538 with a stopping service of five vehicles which may be to Newport, as a three coach formation plus two vehicles of tail traffic on 8 July 1950. The train is above the maximum load of 120tons for the Seven Mile Bank, for which an assistant engine would be needed from Talyllyn. (T.C. Cole)

In the 1920s, the small Prairie 45XXs were a common sight on the Newport to Brecon services. Here Ebbw Jct's 4593 awaits departure with a return service. (W.A. Camwell/SLS)

2247 with the 2.10pm service to Newport standing in the Bay with 2298, off the Bassaleg Goods going to turn on the station turntable on 17 September 1962. (R.E. Toop)

Ebbw Junction's long term 2239 with a two coach service to Newport on 7 May 1954. (R.K. Blencowe)

New Ebbw Junction acquisition 3201 with the 2.5pm to Newport in the bay in September 1962.
(Michael Roach)

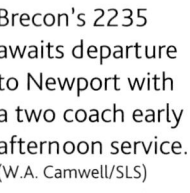

Brecon's 2235 awaits departure to Newport with a two coach early afternoon service.
(W.A. Camwell/SLS)

From the west, the Neath to Brecon service was always worked by Neath Panniers as here with 8732 on the 10.25am from Neath Riverside on 29 September 1962. (W.G. Sumner)

From the mid 1950s, the Standard 2MT 2-6-0s dominated the scene at Brecon and here 46516 has a service to Mid-Wales on 29 September 1962. (W.G. Sumner)

Brecon based 465XXs worked the 2.10pm Brecon to Newport and 6.55pm return until the end of the 1950s as here with 46522 waiting to depart. (R.E. Toop)

Ebbw Junction Panniers were much in evidence on the 8.3am up from Newport until the final day as here with 3747 which has repositioned the ECS for the back working at 12.10pm on 29 September 1962. (W.G. Sumner)

The crew pose for a last time at Brecon as 4679 prepares to work the 12.10pm to Newport on the last day of service 29 December 1962. (E.R. Mountford/Great Western Trust)

4679 awaits departure to Newport on 8 August 1962 with the 6.15pm service. By this time the service was worked exclusively by Panniers and 2251 Class both provided by Ebbw Jct. (Malcolm James)

Above left: Brecon Ely Place Goods yard entrance, closed in September 1955 and seen here in October 1969. (Robert Darlaston)

Above right: A busy scene at Brecon station in the rain on 6 August 1962 with all platforms occupied. Left to Right – the 4pm ex-Hereford arr. 5.56pm with a 465XX, 3pm ex-Newport with coaches through to Neath arr. 5.36pm dep. 6.20pm with 3687, 4.10pm ex-neath arr. 5.51pm forming 6.15pm to Newport to be worked by pannier in siding. (Robert Darlaston)

Below: 3706 arrives at Brecon with a service from Newport in 1962, the view showing the station buildings and SB. (D.K. Jones Collection)

LOCATION ANALYSIS • 183

An LNW 0-6-2T with a Brecon service at Hereford in 1935. (SLS)

Brecon Mountain Railway

Following the closure of the line to Brecon in 1964, a group of enthusiasts set up the narrow-gauge Brecon Mountain Railway running from Pant to Torpantau on the original trackbed and passing through Pontsticill and Dolygaer. Running on a gauge of 1ft 11¾in offered the opportunity to travel through the beautiful scenery of the Brecon Beacons National Park and makes a stop alongside the reservoir at Pontsticill. Vintage steam locomotives work the railway, No. 1 *Santa Teresa*, No. 2 *Baldwin* and No. 3 *Sandy River*.

A Brecon Mountain Railway service running into Pontsticill. This private railway now operates between Pant and Torpantau, just below the site of the former station. (Ken Mumford Presentations)